A Practical Guide to the Histology of the Mouse

A Practical Guide to the Histology of the Mouse

Cheryl L. Scudamore
Mary Lyon Centre, MRC Harwell, UK

WILEY Blackwell

This edition first published 2014 © 2014 by John Wiley & Sons, Ltd
Text illustrations: © Veterinary Path Illustrations, unless stated otherwise

Registered office: John Wiley & Sons, Ltd, The Atrium, Southern Gate, Chichester, West Sussex, PO19 8SQ, UK

Editorial offices: 9600 Garsington Road, Oxford, OX4 2DQ, UK
　　　　　　　　The Atrium, Southern Gate, Chichester, West Sussex, PO19 8SQ, UK
　　　　　　　　111 River Street, Hoboken, NJ 07030-5774, USA

For details of our global editorial offices, for customer services and for information about how to apply for permission to reuse the copyright material in this book please see our website at www.wiley.com/wiley-blackwell.

The right of the author to be identified as the author of this work has been asserted in accordance with the UK Copyright, Designs and Patents Act 1988.

Designations used by companies to distinguish their products are often claimed as trademarks. All brand names and product names used in this book are trade names, service marks, trademarks or registered trademarks of their respective owners. The publisher is not associated with any product or vendor mentioned in this book.

Limit of Liability/Disclaimer of Warranty: While the publisher and author(s) have used their best efforts in preparing this book, they make no representations or warranties with respect to the accuracy or completeness of the contents of this book and specifically disclaim any implied warranties of merchantability or fitness for a particular purpose. It is sold on the understanding that the publisher is not engaged in rendering professional services and neither the publisher nor the author shall be liable for damages arising herefrom. If professional advice or other expert assistance is required, the services of a competent professional should be sought.

Library of Congress Cataloging-in-Publication Data

Scudamore, Cheryl L., author.
  A practical guide to the histology of the mouse / Cheryl L. Scudamore.
      p. ; cm.
  Includes bibliographical references and index.
  ISBN 978-1-119-94120-0 (cloth)
  I.   Title.
  [DNLM: 1. Mice- anatomy & histology. 2. Mice-physiology. 3. Animals, Laboratory. 4.
Histocytological Preparation Techniques. 5. Specimen Handling. QY 60.R6]
  SF407.M5
  616.02′7333 – dc23
　　　　　　　　　　　　　　　　　　　　　　　　　2013028957

A catalogue record for this book is available from the British Library.

Wiley also publishes its books in a variety of electronic formats. Some content that appears in print may not be available in electronic books.

Cover image: © Veterinary Path Illustrations, unless stated otherwise
Cover design by Steve Thompson

Set in 10.5/12.5pt Minion by Laserwords Private Limited, Chennai, India

1 2014

Contents

List of contributors

Elizabeth McInnes
Cerberus Sciences
Thebarton
Australia

Lorna Rasmussen
Cerberus Sciences
Thebarton
Australia

Aude Roulois
GlaxoSmithKline R&D
UK

Cheryl L. Scudamore
Mary Lyon Centre
MRC Harwell
UK

Ian Taylor
Huntingdon Life Sciences
Eye Suffolk
UK

Foreword

Mice are crucial partners in contemporary translational science and may be our most valuable species for genetic modelling of mammalian disease due to the genetic characterization of many inbred and recombinant inbred mice and of increasing numbers of genetically engineered mice (GEM). Recognizing and characterizing deviant phenotypes and developing genetically and pathophysiologically valid models of disease conditions require sophisticated knowledge of mouse anatomy and pathology. A lack of comprehensive and practical contemporary resources on mouse anatomy and pathology and scarcity of experienced mouse pathologists has hindered extensive application of histomorphology and pathology in phenotyping and other biomedical research involving mice and has led to concerns regarding adequate and accurate characterization of mouse disease phenotypes by pathology. *A Practical Guide to the Histology of the Mouse* is an ambitious and comprehensive combination of practical information for the application of mouse histology in research and phenotyping, with detailed and illustrated information on macroscopic and microscopic mouse anatomy. I expect this book to become an essential resource for diverse laboratories and programmes that use mouse models.

Now 40 years after the first transgenic mice, genetic model mice continue to increase, and ongoing international efforts to expand their genetic and phenotypic data and improve access to those data should further enhance their utility as genetic and phenotypic translational research models. The international knockout mouse consortium (IKMC) is creating genetically defined mice that carry mutations in every functional gene, aiming for about 15 000 new lines by 2020. The Collaborative Cross is another international initiative to create genetically diverse recombinant inbred strains, also for genetic modelling. Intentionally outbred mice offer additional genetic modelling options. Phenotyping of these genetic models provides the key to understanding gene functions in a complex mammalian system. The international mouse phenotyping consortium (IMPC) is conducting baseline phenotyping on all of the IKMC-generated mice, to enhance their utility to the global research community. The mouse phenome database (MPD) at the Jackson Laboratory is another important internationally contributed phenotype resource. The application of histopathology in these initiatives, to date, has been limited. This book will facilitate efforts to improve application of histopathology in research involving mice for (i) troubleshooting unexpected morbidity or mortality in research projects, (ii) characterizing disease phenotypes, and (iii) validating genetic models.

A Practical Guide to the Histology of the Mouse presents each organ system in the context of its development, its anatomic structures and functions, with specimen collection and sampling options, and common strain-, sex-, or age-related expected (background) findings. The level of detail, excellent illustrative images and insights on common artefacts and misinterpretations will be appreciated by experienced pathologists as well as those who are new to mouse pathology, or to research-specific histomorphology assessments in mice. A unique feature of this book is the valuable guidance for study design, handling specimens, records and data, which provides important insights for optimizing pathology results and pathology data in any biomedical research setting.

The editor and other contributing authors are highly experienced, published authorities in rodent pathology. They provide a wonderfully detailed, systematic and comprehensive approach that reflects their pathology expertise and their extensive experience in discovery, safety, toxicity and carcinogenicity studies. I expect this book to be widely used and recommended for explaining and teaching mouse anatomy and pathology for many years and to be one of the most useful mouse resources in many research laboratories.

Cory Brayton
Director, Phenotyping Core
Associate Professor, Molecular and Comparative
Pathobiology
Johns Hopkins University School of Medicine

Preface

Mouse models of disease are used extensively in biomedical research with many hundreds of new models being generated each year. In addition, global consortia are working to knock out every gene in the mouse genome. Full phenotypic analysis of mouse models requires the morphologic analysis of tissue samples, which, in turn, requires an understanding of the basic histology and common background variations in the histology of mice. However there is a global shortage of pathologists trained to evaluate mouse tissues and many researchers have to attempt this analysis themselves. While it would be most appropriate to have guidance from an experienced pathologist, in reality this is not always possible.

This book is therefore aimed at veterinary and medical pathologists who are unfamiliar with mouse tissues, and scientists who wish to evaluate their own mouse models. It aims to provide practical guidance on the collection, sampling and analysis of mouse tissue samples in order to maximize the information that can be gained from these tissues. Getting the most information from each individual animal is good practice scientifically, financially and from a welfare point of view, as a contribution, by reduction, to the 3Rs.

As well as illustrating the normal microscopic anatomy of the mouse, some of the common anatomic variations, artefacts associated with tissue collection and background lesions are described and explained to help the observer distinguish these changes from experimentally-induced lesions. Methods for recording and analysing the data gained from the pathological analyses are also described. Histology is a relatively old discipline and there are many adjunct techniques (for example special stains and immunohistochemistry) over and above the standard haematoxylin and eosin (H&E) stain, which can be used to add value to the information gained from fixed tissues. In addition new histological techniques are emerging all the time. It is not possible to cover all these techniques comprehensively in this book but examples have been given where appropriate. There are also often many alternative methods for achieving similar end points and this book does not aim to be comprehensive but rather to present tried and tested methods that the authors know to work, particularly in a high throughput environment.

Inevitably, in a book of this kind there will be some omissions, some points for debate and some errors. The first two are unfortunate consequences of a book where space is limited and the nature of pathology is subjective, but we apologize for the errors. The authors would welcome constructive comments and suggestions for future editions. Readers may find some additional books helpful to fill in some of the omissions. A basic primer in mammalian histology, of which there are many, will be useful for complete novices and a recently published, comparative atlas of mouse and human histology (Treuting and Dintsiz 2011) is recommended for more detailed coverage. Although this book attempts to point out some of the common artefacts and background findings in the mouse, it is not possible to cover all of the common spontaneous pathology seen in mouse strains and additional books will also be useful in a library (Maronpot 1999; McInnes 2012).

It is important to acknowledge that this book has come about from the knowledge distilled from the numerous histologist and pathologist colleagues with whom we have worked with over the years. These colleagues are too numerous to mention individually but they know who they are and we thank them for their training, advice and wisdom. One person who does deserve a special mention is the artist from Veterinary Path Illustrations who produced the beautiful drawings for this book. Finally we must thank our current colleagues, friends and family for supporting us in this effort over the last few years.

Cheryl L. Scudamore
Liz McInnes
Aude Roulois
Ian Taylor

References

Maronpot, R. (1999) *Pathology of the Mouse. A Reference and Atlas*, Cache River Press, Saint Louis, MO.

McInnes, E.F. (2012) *Background Lesions in Laboratory Animals – A Color Atlas*, Saunders, Elsevier, Edinburgh.

Treuting, P. and Dintsiz, S.M. (2011) *Comparative Anatomy and Histology. A Mouse and Human Atlas*, Academic Press, London.

About the companion website

This book is accompanied by a companion website:

www.wiley.com/go/scudamore/mousehistology

The website includes:
- Powerpoints of all figures from the book for downloading
- PDFs of tables from the book for downloading

Chapter 1
Necropsy of the mouse

Lorna Rasmussen and Elizabeth McInnes
Cerberus Sciences, Thebarton, Australia

Necropsies on mice are a fundamental part of the research process (Fiete and Slaoui 2011) and it is vital that, in every laboratory where animal research is conducted, prosectors (persons who perform necropsies or post mortem examinations) are trained to perform a complete mouse necropsy.

Necropsies on mice are performed for a number of reasons including harvesting of tissues for research, health surveillance and investigation of disease. The process may involve collection of tissues for pathology (e.g. for phenotyping or analysis of research models), but also collection of appropriate samples for microbiological and parasitological examinations for disease identification. Autolysis after death begins immediately after the onset of hypoxia as a result of cessation of blood flow (Slauson and Cooper 2000). Autolysis of the small intestine will commence within 10 minutes of the death of the animal, resulting in the swelling of villus tips and epithelial denudation of the villi (Pearson and Logan 1978a,b). Bone marrow and adrenals are also susceptible to rapid autolysis. Storage of mouse carcases in a refrigerator (2–4 °C) is recommended to avoid rapid autolysis, which may occur in the warm atmosphere of an animal house.

A systematic approach to the mouse necropsy, which allows the examination of all the tissues in the animal in the most expedient manner, is recommended (Slaoui and Fiette 2011). In this chapter, the authors describe a recommended protocol, but variations on this method may exist, depending on the target organ or disease model. It is important to conduct a complete necropsy and to avoid the temptation of just looking at the organs of interest or selection of organs (Seymour *et al.* 2004). Necropsy results should always be viewed in conjunction with ante mortem clinical signs and haematology and biochemistry results. It is advisable to prepare all instruments, sample collection materials, camera and forms before beginning the necropsy (Knoblaugh *et al.* 2011). Necropsy personnel should always have access to the experimental study plan so that they can collect particular tissues that are pertinent to the study (Fiette and Slaoui 2011).

Mouse necropsies carry risks of zoonoses, allergen exposure and exposure to hazardous materials

A Practical Guide to the Histology of the Mouse, First Edition. Cheryl L. Scudamore.
© 2014 John Wiley & Sons, Ltd. Illustrations © Veterinary Path Illustrations, unless stated otherwise.
Published 2014 by John Wiley & Sons, Ltd. Companion Website: www.wiley.com/go/scudamore/mousehistology

such as formalin (Fiette and Slaoui 2011). Appropriate equipment is thus necessary to conduct the mouse necropsy procedure. This includes equipment to conduct appropriate and ethical methods of euthanasia, if the mouse is still alive. Different methods of euthanasia of laboratory mice include carbon dioxide asphyxiation, barbiturate overdose and cervical dislocation (Seymour *et al.* 2004). Carbon dioxide asphyxiation is a rapid and humane form of euthanasia for mice over the age of seven days (Seymour *et al.* 2004), but can result in significant agonal haemorrhage, which can complicate microscopic examination of the lungs. Only one mouse should be placed in the perspex container at a time and carbon dioxide gas slowly added to the chamber. A flow rate of 20% V/min CO_2 as a gradual fill or slow filling method for the chamber results in least evidence of stress in mice (Valentine *et al.* 2012). Barbiturate overdose is an effective and efficient form of euthanasia and requires the use of pentobarbital sodium (Seymour *et al.* 2004). Decapitation of adult mice should be avoided because the method has welfare concerns and may not be accepted by the ethics committee of an institute or the Home Office (United Kingdom) (Seymour *et al.* 2004). Cervical dislocation was first approved for mouse euthanasia in 1972 by the AVMA Panel on Euthanasia (Carbone *et al.* 2012). The disadvantages of cervical dislocation are that although it may be a quick and efficient method of euthanasia, it causes damage to the tissues in the cervical area and may cause the release of large amounts of blood into the body cavities (Seymour *et al.* 2004). In addition, Carbone and co-workers (2012) examined spinal dislocation and noted that of the 81 mice that underwent cervical dislocation, 17 (21%) continued to breathe and euthanasia was scored as unsuccessful.

Further equipment for mouse necropsy includes a controlled air-flow cabinet or down draft table and plastic containers of formalin and syringes. The controlled air-flow cabinet is not always available in all facilities and is not essential, however it does reduce the risk of noxious substance inhalation (e.g. formalin), or allergen exposure and it is also essential to control the spread of known pathogens or zoonotic agents from the mouse carcase. Cover slips and glass slides may also be required for the preparation of cytology and parasitology specimens. It is important to have a flat and contained area in which to perform the necropsy. A flat board made of rubber or plastic is advisable so that it can be decontaminated or autoclaved if necessary. A metal tray to hold the mouse carcase may also be useful. Furthermore, two pairs of forceps, scalpel blades and scalpel blade holder, one pair of sharp-edged scissors, disinfectant spray and racks for Eppendorf containers and other plastic containers as necessary depending on the sampling protocol will be required. In addition, paper towels are necessary throughout the necropsy procedure (Knoblaugh *et al.* 2011). Some prosectors will prefer to pin the mouse carcase to a flat cork board while others may prefer to move the mouse during the necropsy procedure. A metric ruler is important for measuring organs and lesions (Knoblaugh *et al.* 2011). Plastic containers of 10% neutral buffered formalin (NBF) or 4% paraformaldehyde as well as containers of other fixatives (such as modified Davidson's for the fixation of testes and eyes) and containers for microbiology and molecular biology samples should be present at the start of the necropsy process and should all be labelled with the correct mouse identification number. A syringe and plastic cannula or needle (22G) for perfusing the lungs with formalin should also be present at the start (Braber *et al.* 2010). Mouse adrenals, pituitary gland and lymph nodes are notoriously small and difficult to handle and the use of cassettes with foam pads or biopsy bags to store them in so that they are not mislaid is highly recommended (Knoblaugh *et al.* 2011).

1.1   Recording of findings

During the necropsy the prosector must record, in some form, all the observations made during the necropsy examination. This will provide a valuable aid at the end of the necropsy and after the histopathological examination of the slides has been conducted, to form conclusions about what abnormalities were observed and may form part of the data set for the experimental group (Chapter 2).

The observations may be written down on a specific form designed for that purpose or they may be entered into a computer data-collection program. It is a good idea to develop a checklist for the mouse necropsy procedure, which is referred to each time and on which organ systems may be ticked off as they are inspected and collected.

The correct identification of the animal is very important. The researcher must examine the information on the cage lid, the ear tag of the animal or the ear notches (Figure 1.1) or scan the microchip in order to confirm the exact identification of the animal to be necropsied. The prosector may have to consult a specific key indicating what number each ear notch represents as ear notch keys are usually specific to particular research institutions. It is also important to label all samples generated from the necropsy with the same mouse identification number. The age, sex, strain, genotype, reason for submission and study number (if appropriate) should also be recorded if they are known (Seymour *et al.* 2012) the body weight, in grams, at the time of necropsy should be recorded. Retaining the identification (ear tag, ears or chip) with the fixed organs is good practice and acts as a safeguard to ensure that organs can be accurately identified if the external labelling becomes damaged.

During the necropsy process, the prosector should make a note of the characteristics of the abnormalities observed. The abnormal organs should be identified and information about the size, site, shape (for example wedge-shaped lesion or rounded edges of organ), colour (see Chapter 2), the consistency (whether the organs are hard or soft to the touch) and borders (sharp demarcation between normal and abnormal tissue or diffuse borders) of the lesion should be recorded. In addition, the appearance of the cut surface of the abnormality should be described and the normal or abnormal contents of some of the hollow organs such as the urinary bladder and the small and large intestines should be mentioned. Information on whether the lesion is focal, multifocal, focally disseminate or diffuse should be included.

1.2   Bleeding technique

There are a number of efficient methods for collecting blood from mice (Hoff 2000). If blood samples are required at necropsy, cardiopuncture for blood collection may be performed by inserting the needle 2 mm right of the xiphoid bone to the level of hub and gently withdrawing the plunger (Figure 1.2) but this method can cause artefactual increases in enzymes due to damage to the cardiac muscle. Retro-orbital bleeding under anaesthesia and tail tip amputation may also be used to collect blood (Seymour 2004). Bleeding may also be performed via the anterior thoracic aperture (Frankenberg 1979). A 1 ml syringe and 23 G needle should be used and the blood should be transferred quickly to anticoagulant treated plastic 1 ml containers (for haematology) or nontreated plastic 1 ml containers for the production of serum from the clotted blood. Proficient prosectors should be able to collect between 0.6 to 0.8 ml of blood. Some workers have recommended exsanguination from the abdominal aorta (Fiette and Slaoui 2011).

1.3   Perfusion

Perfusion is recommended for certain indications to ensure minimal autolytic change and to maximize morphology and retention of antigens in the tissues.

Figure 1.1   Ear notch used to identify mouse.

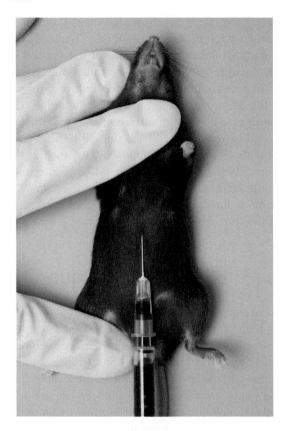

Figure 1.2   If blood samples are required, cardiopuncture may be performed.

The decision to perfuse animals should be made after a cost/benefit analysis of the procedure. Perfusion requires more time, usually more equipment and, if not performed carefully by experienced personnel, can create more artefacts than it prevents (Chapter 4). Protocols also have to be followed carefully to ensure perfusion is completed in a humane manner (Seymour *et al.* 2004). There are a number of options for perfusion, which require varying amounts of additional equipment (Hayat 2000). The following describes a simple technique that can be used with minimal additional equipment. Briefly, the prosector should prepare two 10 ml syringes with one containing saline and the other containing the fixative of choice (usually 10% neutral buffered formalin). The mouse should be anaesthetized using a peritoneal injection of pentobarbitone (0.1 ml/10 g

body weight) (Seymour *et al.* 2004). The jugular veins should then be exposed below the salivary glands and the thoracic cavity should be opened. The jugular veins are then cut and the needle of the syringe containing saline should be inserted into the left ventricle of heart. The heart should be perfused with saline and saline should soon be visible exiting from the severed jugular veins. After injecting 4–8 ml of saline, the procedure should be repeated with the fixative. The organs will stiffen and become grey in colour (Seymour *et al.* 2004).

1.4   External examination

The external examination is the first procedure to carry out on the animal's body. The prosector should examine the animal's general condition, whether it is obese, emaciated or normal and to establish whether there is evidence of skeletal muscle atrophy (Fiette and Slaoui 2011). Overgrooming or barbering and whisker plucking is common in C57BL6/J mice and may be evidence of dominance behaviour in cage mates (Sarna *et al.* 2000). It is also important to look for the presence of skin ulceration (Figure 1.3), loss or abnormalities of fur or any superficial lesions. The presence of a rough, dry, scaly skin may indicate the presence of parasites. It is advisable to examine the external openings – that

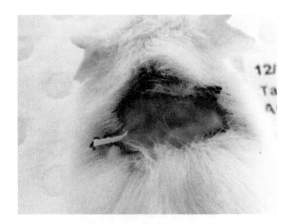

Figure 1.3   Focal ulcerative lesion in the skin noted in scapula region of mouse.

is the eye, ear, mouth and urogenital orifices – for the presence of blood or discharge. Mice of the C57BL/6 strain are susceptible to ulcerative dermatitis (severe skin lesions) and may show an incidence of greater than 20% (Sundberg and King 1996). If phenotyping of the mouse is required, the prosector should always collect a skin sample from the same area of the body such as the thorax (Seymour *et al.* 2004) and the skin should be placed on cardboard before being placed in 10% neutral buffered formalin to avoid curling and distortion.

Examination of the skin should include a search for the presence of traumatic wounds (which are common in some strains of group-housed male mice) including abscesses of the face and retrobulbar region, which are common in mice and are generally caused by bite wounds becoming infected with *Staphylococcus aureus* (Clarke *et al.* 1978). Distension of the abdomen and the presence of skin or mammary gland tumours should also be noted.

Examination of the inside of the mouth is important and the prosector should make a careful note of the state of the tongue, the oral mucosa, the lips and upper teeth. White mucous membranes in the mouth, may indicate anaemia. Small haemorrhages of the mucosa may be noted on the gums and this may indicate the presence of an infectious disease or toxaemia. The prosector should also examine the oral mucosae for the presence of ulcers and blisters or vesicles. Abnormalities of the teeth include loss, overgrowth, erosion, discolouration and fractures and are all common in mice (Figure 1.4). Severe emaciation in the mouse is often linked to dental abnormalities. The presence of blood at the nares or nostrils is important and should be noted and may indicate rupture of a wall of a blood vessel within the pulmonary system.

Cataracts (opacification of the crystalline lens of the eye) (Figure 1.5) are observed in up to 25% of Swiss CD-1 mice by 28 months of age (Taradach and Greaves 1984). This lesion may be seen at necropsy and is characterisation by the cloudy-white colour of the eye. Small eyes i.e. microophthalmia (Figure 1.6) and anophthalmia are noted commonly in C57BL/6 mice (Smith *et al.* 1994). The Harderian gland is a bilobular pink, horseshoe-shaped gland located in

Figure 1.4   Overgrown and misaligned incisor teeth in the mouse.

Figure 1.5   A cataract in the right eye is characterisation by opacification of the crystalline lens of the eye.

the orbit of the eye of all nonprimate vertebrates and in mice this gland characteristically produces high concentrations of porphyrin (brown pigment) under hormonal control (Margolis 1971). Genetically different strains of mice manifest different amounts of porphyrin in their Harderian glands (Margolis 1971). There are marked sex differences in the Harderian gland of the C3H/He strain of mice. Female (but not male) glands contain large amounts of porphyrin (Shirama *et al.* 1981). In addition, the prepuce should be examined for the presence of inflammation, purulent material and penile prolapse, or ulceration. The scrotum should

Figure 1.6   Micro-ophthalmia in the right eye of a C57 black mouse.

Figure 1.7   Rectal prolapse in the mouse may be observed in the perineal area.

be examined for skin lesions and enlargement and the vulva should be examined for haemorrhage and purulent discharges.

It is important to examine the perineum, which is the skin adjacent to the rectum. If the mouse has suffered from diarrhoea, the perineal area often contains small flecks of faeces, which may be blood-stained (Sundberg *et al.* 1994). Rectal prolapse can also be observed in this area (Figure 1.7). Rectal prolapse is fairly common in mice and is characterized by the presence of intestinal tissue at the rectum, often with ulceration and infection (*Helicobacter spp.* are a potential cause of rectal prolapse).

1.5   Weighing of organs

Various authors have published recommendations on the weighing of organs at necropsy (Sellers *et al.* 2007; Michael *et al.* 2007). In all cases, the organs should be weighed free of surrounding fat and connective tissues (Fiette and Slaoui 2011).

1.6   Positioning of mouse for necropsy and removing the skin

The mouse is placed on its back after spraying the abdominal surface with 95% ethanol or a disinfectant spray. The mouse limbs may be pinned to a cork board or left unpinned, as preferred by the prosector. A small cut is made at the level of the pubis using a scalpel blade or scissors, and then a longitudinal cut is made along the central midline, through to the chin. The skin is then dissected away from the body leaving the abdominal wall intact (Figure 1.8). The subcutis, superficial lymph nodes, mammary glands, penis and skeletal muscles will be apparent and should be examined at this point.

The subcutaneous tissues may display gelatinous fluid and this indicates the presence of widespread oedema (anasarca). This condition may be observed

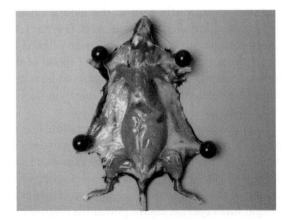

Figure 1.8   The mouse is placed on its back and a longitudinal incision is made from chin to pubis before dissecting the skin away from the body leaving the abdominal wall intact.

in cases of severe chronic renal, heart or liver failure. At this point in the necropsy procedure, it is advisable to look for generalised colour changes such as pale, white subcutaneous tissues, which may indicate widespread anaemia or yellow tissues which indicate the presence of jaundice or icterus. The lymph nodes are small, bean-shaped structures and are not easy to locate in a healthy mouse. Murine lymph nodes may become haemorrhagic and cystic with age. The position of the most commonly harvested lymph nodes in the mouse are indicated in Figure 1.9 and

described by Van den Broek *et al.* (2006). Lymph nodes may become enlarged due to inflammatory processes in organ systems or due to the presence of tumours (Figure 1.10). In general, the cervical lymph nodes are situated above the submandibular salivary glands, the axillary lymph nodes are present in the axillary fossa of the forelimbs and the inguinal lymph nodes are present in the fat pad in the fossa of the hind limbs – see Vincenzo Covelli's *Guide to the Necropsy of the Mouse*, http://eulep.pdn.cam.ac.uk/Necropsy_of_the_Mouse/printable.php (accessed 17 July 2013) and Figure 1.11.

The salivary glands (submandibular, sublingual and parotid) are paired organs found in association with the paired submandibular lymph nodes in the region below the chin and adjacent to the larynx and

Figure 1.9 The position of the mandibular (A), deep cervical (B), superficial parotid (C), axillary (D), accessory axillary (E), tracheobronchial (F), caudal mediastinal (G), gastric (H), renal (I), mesenteric (J), inguinal/subiliac (K), medial iliac (L) and popliteal (M) lymph nodes is indicated on the diagram.

Figure 1.10 Enlarged lymph nodes as a result of lymphoma.

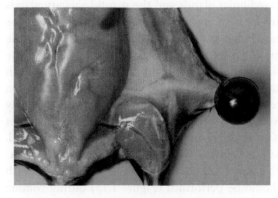

Figure 1.11 Inguinal lymph node in fat pad in the inguinal area.

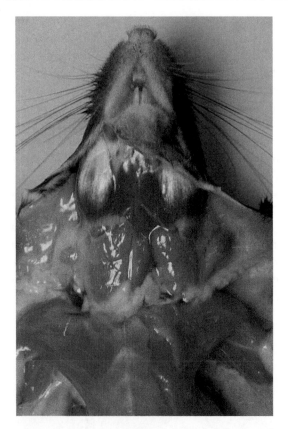

Figure 1.12 The salivary glands are paired organs found in association with submandibular lymph nodes below the chin.

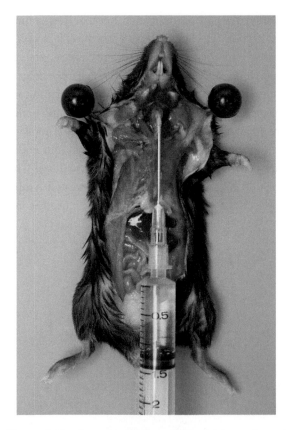

Figure 1.13 A nasotracheal wash should be performed during routine health monitoring.

cranial trachea (Figure 1.12). The submandibular salivary glands are large and are situated in the central region of the neck (Fiette and Slaoui 2011). The sublingual glands are situated above the submandibular glands and the parotid salivary glands are located laterally and extend to the base of the ear (Fiette and Slaoui 2011). In intact adult male mice, the submandibular salivary gland is twice the size of the salivary gland in female animals (Frith and Ward 1988). Once the salivary glands have been removed, the trachea is exposed. The prosector may wish to perform a nasotracheal wash at this point if an upper respiratory tract infection is suspected. Scissors or forceps should be used to raise the trachea for a nasotracheal wash. A nasotracheal wash should be performed (Figure 1.13) during routine mouse health monitoring and also if infection is suspected in the lungs or upper respiratory tract.

A detailed method for performing a nasotracheal wash is described in McInnes *et al.* (2011).

There are five pairs of mammary glands in the female mouse and these are situated within fat pads, laterally, on both sides of the animal, immediately in front of and behind the forelimbs and immediately cranial to the hindlimbs and in the inguinal canal region (Figure 1.14). The mammary glands are easy to see when the female mouse is lactating (Knoblaugh *et al.* 2011). When the mammary glands are completely developed, they may extend through to almost all of the subcutaneous tissues apart from some areas of the back (Fiette and Slaoui 2011) (Figure 1.15). Mammary tumours that manifest as hardened or cystic masses may be observed at a high incidence in some strains of mice.

The clitoral glands in the female animal and the preputial glands in the male mouse should

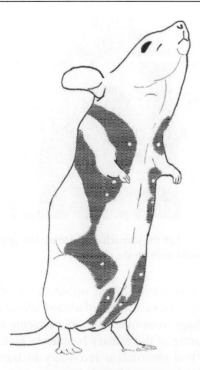

Figure 1.14 Location of mammary tissue in the female mouse. There are five pairs of mammary glands in the female mouse and these are situated within fat pads, laterally, on both sides of the animal, immediately in front of and behind the forelimbs and immediately cranial to the hindlimbs and in the inguinal region.

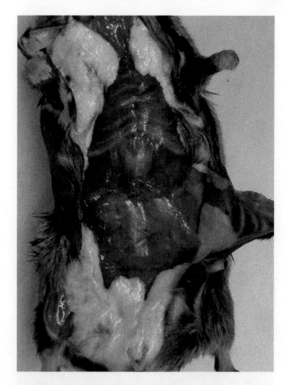

Figure 1.15 Lactating mammary tissue may extend to almost all of the subcutaneous tissues.

be examined at this point. Clitoral or preputial glands are modified sebaceous glands, which are leaf-shaped, grey in colour and soft in consistency (Slaoui and Fiette 2011). Abscessation of the preputial glands is common in male mice (Suwa *et al.* 2001) (Figure 1.16). The preputial glands are situated between the penis and rectum in the male animal. In the female mouse, the clitoral glands are found in the same region, cranial to the vulva.

1.7   Opening the abdominal cavity and exposing organs

The abdominal cavity is opened by grasping a small piece of muscle in the midline of the abdomen with the forceps. The muscle is then lifted and a small

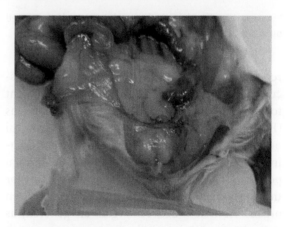

Figure 1.16 Abscessation of the preputial glands is common in the male mouse.

incision is made with the scissors. One blade of the scissors is then inserted into the abdominal cavity and the incision is continued along the midline to the xiphoid bone cranially and the pubis caudally. It is advisable to split the pelvis in two in order to expose the reproductive organs,

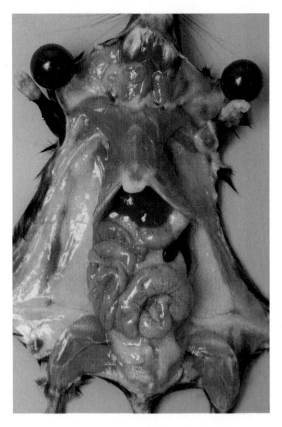

Figure 1.17   The abdominal musculature may be sectioned on either side to expose the abdominal organs.

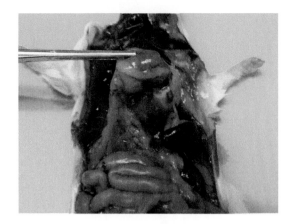

Figure 1.18   Lift the stomach and examine the spleen and pancreas found beneath the stomach.

rectum and urinary bladder. The abdominal musculature may be sectioned on either side, adjacent to the ribs, in order to expose the abdominal organs (Figure 1.17). Before removing any abdominal organs, the prosector should examine the abdominal cavity for the presence of organs in unusual positions, distension of the intestinal tract and for the presence of fluids, blood or adhesions between organs. Adhesions between organs are generally as a result of a previous inflammatory process such as peritonitis. Acute peritonitis tends to manifest with the presence of a yellow, often cloudy fluid with occasional strands of nonadherent, yellow material (fibrin) present. The presence of clear fluid in the abdominal cavity may indicate either kidney, heart or liver failure and may also be observed in cases of lymphoid tumours. The presence of

blood or bloodstained fluid within the abdominal cavity may indicate rupture of a major blood vessel, haemorrhage from ovarian cysts in ageing female mice, rupture of the urinary bladder in male mice with urethral obstruction secondary to traumatic damage to the external genitalia or rupture of the spleen due to trauma.

After the initial viewing of the exposed abdominal organs, it is advisable to lift the stomach and to examine the spleen and pancreas found beneath the stomach (Figure 1.18). The spleen in the mouse is generally shaped as an elongated oval and is connected to the stomach. The spleen in immunodeficient mice may be very small in comparison with that of an immunocompetent mouse. The spleen should be grasped with forceps and removed. Occasionally ectopic splenic tissue may be observed in the mesentery of the gastrointestinal tract. Black mice of the C57BL/6 strain may demonstrate a blackened portion of the spleen (Figure 1.19) due to the presence of melanin in the red pulp (Suttie 2006). In the mouse, enlargement of the spleen (splenomegaly) is observed commonly and may indicate a tumour process (commonly lymphoma, but other tumour types also occur e.g. haemangiosarcoma), lymphoid hyperplasia or an increase in extramedullary haemopoietic cells (granulocytes or precursor erythrocytes) in response to systemic inflammatory processes or red blood cells (in response to anaemia) in the red pulp (Figure 1.20).

Figure 1.19   The mouse spleen is generally an elongated oval and in black mice of the C57BL/6 strain may demonstrate focal areas of darkening.

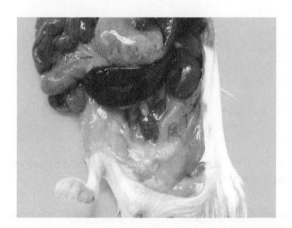

Figure 1.20   Splenomegaly is observed commonly and may indicate a tumour process or an increase in immature haemopoietic cells.

The pancreas is located throughout the mesentery adjacent to the cranial duodenum and may be removed separately or be left attached to the duodenum. The pancreas is a pink, diffuse tissue and is covered with the mesentery. The pancreas may display oedema (fluid present within the interstitium of the pancreas) (Figure 1.21) or atrophy.

The mouse liver is made up of four lobes, right, left, medial and caudate (Knoblaugh *et al.* 2011) (Figure 1.22). It is advisable to remove the entire liver and gallbladder from the abdominal cavity

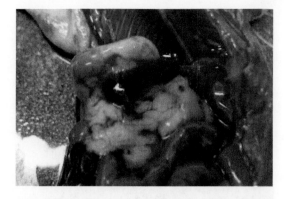

Figure 1.21   The pancreas may occasionally display oedema (fluid present within the interstitium of the pancreas).

Figure 1.22   The mouse liver is made up of four lobes, right, left, medial and caudate with the gall bladder present between the medial lobes.

to fix it whole. The normal colour of the liver is reddish brown. The liver should be examined for abnormal colour, a firm or soft consistency (a firm consistency is often indicative of scarring or fibrosis) and enlargement with rounded edges. Common lesions in the livers of older mice include a diffuse, yellow colour (fatty or lipid vacuolation of the hepatocytes), enlargement, a pitted, granular surface (generally an indication of inflammation) (Figure 1.23) and tumour masses (Figure 1.24). The gallbladder is present between the medial lobes and is generally filled with a watery, yellow fluid. A distended gall bladder may indicate anorexia and prolonged illness. Lesions in the gallbladder include

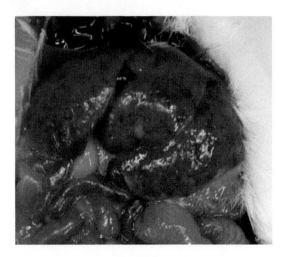

Figure 1.23   Enlargement and a pitted, granular surface of the liver are generally an indication of inflammation.

Figure 1.24   Mass in liver.

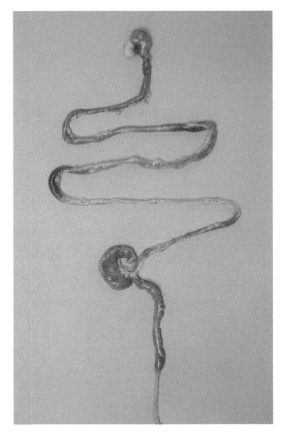

Figure 1.25   The gastrointestinal tract should be laid out for inspection.

severe dilatation, which is a common change in older mice (Lewis 1984).

The gastrointestinal tract should be removed *in situ* from the abdominal cavity. This involves cutting through the oesophagus cranially, close to the insertion of the oesophagus into the stomach and cutting through the rectum caudally and gently separating the intestines from their attachments to the mesentery of the peritoneal cavity. The gastrointestinal tract should then be laid out for inspection (Figure 1.25). The prosector should examine the serosal intestinal surfaces and the intestinal contents carefully looking for evidence of dilatation, haemorrhagic and liquid contents (particularly in the colon and rectum), thickening of the intestinal walls and rupture of the tract and possible adhesions. Merkel's diverticulum is a congenital defect which causes an out pouching of the small intestine generally in the region of the jejunum (Figure 1.26). The excessive accumulation of gas in the gastrointestinal tracts resulting in severe gastric and intestinal dilatation (Figure 1.27) may be caused by obstruction of the nasal passages (Nakajima and Ohi 1977). In these cases, histopathological examination of the nasal turbinates often results in the diagnosis of a severe purulent rhinitis.

The larger mesenteric lymph nodes (situated adjacent to the junction between the caecum and colon) (Figure 1.28) should be inspected for

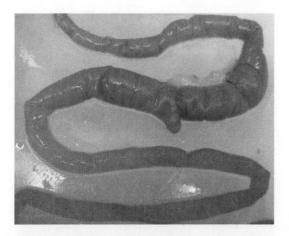

Figure 1.26   Merkel's diverticulum is a congenital defect which causes an out pouching of the small intestine generally in the region of the jejunum.

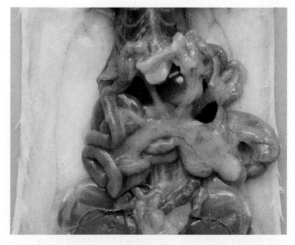

Figure 1.28   The larger mesenteric lymph nodes situated adjacent to the junction between the caecum and colon.

Figure 1.27   Excessive accumulation of gas in the gastrointestinal tract of a mouse suffering from purulent rhinitis.

enlargement and colour changes. The mesenteric lymph nodes may demonstrate the presence of lymphoma, which is very common in certain strains of mice. There is an incidence of up to 50% of lymphoma in female animals and 22% in male animals of the CD1 strain of Swiss mice (Son 2003). Phosphate-buffered saline may be used for the rinsing of the gastrointestinal tract before fixation (Knoblaugh *et al.* 2011). If a faecal flotation is to be performed later, flushing of intestinal contents with formalin (thus enabling faecal contents to be stored) using a 10 ml syringe and a 21G needle is advised.

The entire stomach and intestinal tract may be fixed whole, or the stomach and small and large intestines may be opened and then the entire intestinal tract laid out on cardboard (serosa side down) and fixed to prevent distortion. The prosector may also wish to inflate the intestines with formalin before fixation. Swiss rolls of intestine are created by rolling the inflated intestine in concentric centrifugal circles on piece of cardboard (Moolenbeek and Ruitenberg 1981). If the intestines are opened before fixation, the prosector should examine the intestines paying careful attention to the contents, the colour of the mucosa, the thickness of the wall and the presence of ulceration and perforation. The intestinal wall and mucosa will display multifocal, raised nodules at various points through the small and large intestine. These firm, yellowish-white nodules are Peyer's patches or aggregates of lymphoid tissue and may appear quite prominent (Figure 1.29) and are often mistaken for ulceration. In the case of lymphoma, the Peyer's patches may be enlarged.

The organs of the urinary system consist of the paired kidneys, urinary bladder, paired ureters and urethra. The kidneys are situated on either side of the vertebral column and will be observed once the abdominal organs have been removed (Figure 1.30). The adrenal glands are present at the cranial pole of each kidney (Figure 1.30). The adrenals of female

Figure 1.29   The firm, yellowish-white nodules are prominent Peyer's patches or aggregates of lymphoid tissue.

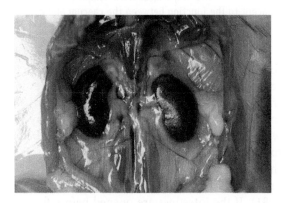

Figure 1.30   The kidneys in the mouse are situated on either side of the vertebral column and the adrenal glands are present at the cranial pole of each kidney.

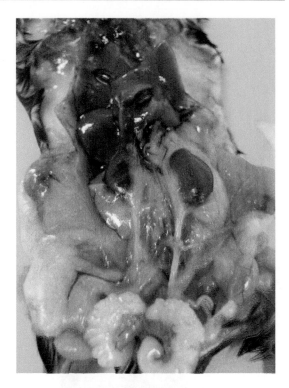

Figure 1.31   The ureters are extremely small, but resilient and may be sampled by lifting the adipose tissue on either sides of the vertebral column.

mice are larger than those of male mice and are generally a pale yellow colour (in contrast to adrenal glands in the male, which are a reddish brown colour). The right kidney is always situated in a slightly more cranial position than the left kidney. The kidneys should be removed. It may be easier to remove the adrenal glands with the kidneys and to fix them together, however it is preferable to fix the adrenals separately using biopsy bags (Knoblaugh *et al.* 2011). It may be useful to use a scalpel to make a longitudinal section in the left kidney and a transverse section in the right kidney in order to distinguish left from right later in the processing procedure (Knoblaugh *et al.* 2011). The ureters are extremely small, but resilient and may be visualized by lifting the adipose tissue on either sides of the vertebral column (Figure 1.31). The kidneys are normally a reddish-brown colour and abnormalities of the kidneys include small, shrunken, firm, pale-yellow kidneys with a roughened, granular surface (Figure 1.32). This may indicate chronic glomerulonephropathy, which is common in older mice (Frith and Ward 1988). Tumour masses such as lymphoma may also be present in the kidneys. Hydronephrosis or pelvic dilatation (Figure 1.33) is a common finding in mice (Goto *et al.* 1984) and is characterized by a large, fluid-filled cyst in the normal position of the kidney and the attached ureter may also be distended with fluid. The kidneys from male animals are larger than the kidneys from female mice. Some strains of ageing C57BL/6 mice are reported to demonstrate a high incidence of chronic progressive nephropathy in both male and female mice (Zurcher *et al.* 1982).

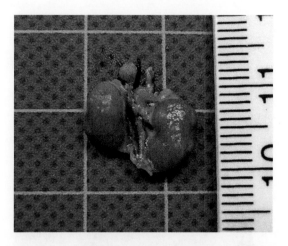

Figure 1.32   Small, shrunken, firm, pale-yellow kidneys with a roughened granular surface.

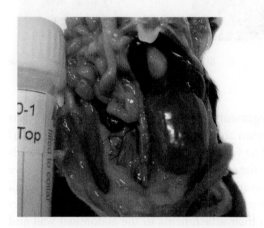

Figure 1.33   Hydronephrosis in the mouse kidney.

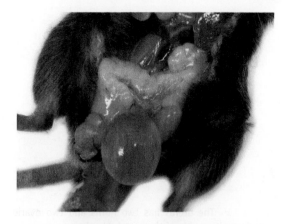

Figure 1.34   Distended urinary bladder in the mouse.

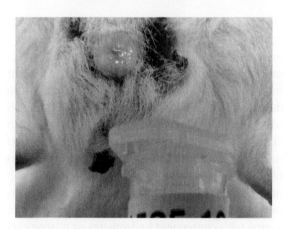

Figure 1.35   Swelling of the prepuce in the male mouse.

The urinary bladder is situated above the prostate gland in the male animal and above and attached to the vagina in the female animal. When removing the urinary bladder, it is important to look for evidence of distension, which may indicate a blockage in the urinary system. In some cases, it may be easier to remove the urinary bladder while it is still attached to the male and female reproductive organs. A urine sample can be taken from the urinary bladder using a 23G needle and 1 ml syringe, if the urinary bladder is full. Urinary bladder distension (Figure 1.34) is a common finding in the mouse and may be related

to uroliths or inflammation of the penis or prepuce and is often secondary to traumatic damage in group housed males (Figure 1.35). The prosector may wish to inflate the urinary bladder with formalin using a syringe and needle and to ligate the urethra to prevent the creation of artefactual folds.

The female reproductive system consists of the uterus, the paired ovaries, the oviducts and the cervix and vagina. The uterus has two horns (Figure 1.36), which are situated on either side of the vertebral column and which arise from the cervix, which is generally below the urinary bladder. The ovaries are small, round organs attached to the tips of the uterine horns and also attached to the abdominal cavity by the mesovarium – see Vincenzo

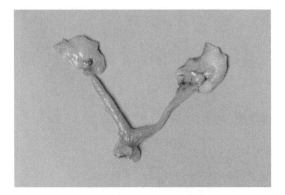

Figure 1.36   The uterus has two horns and two ovaries situated on either side of the vertebral column.

Covelli's *Guide to the Necropsy of the Mouse*, http://eulep.pdn.cam.ac.uk/Necropsy_of_the_Mouse/printable.php (accessed 17 July 2013). The ovaries may display multifocal nodules which represent follicles or corpora lutea. The ovaries may also demonstrate tumours and cysts (Figure 1.37), which will cause enlargement. The oviducts are too small to visualize but connect the uterine tip to the ovary. At necropsy, it is common to see dilated uterine horns filled with turbid fluid (Figure 1.38). Slight dilatation may be due to cyclical change related to oestrous but massive dilatation can be related to imperforate vagina with consequent hydrometra or mucometra which has been seen in 7% of virgin female BALB/c mice (Sheldon *et al.* 1980). In older mice

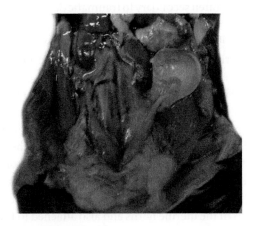

Figure 1.37   Cystic left ovary in the female mouse.

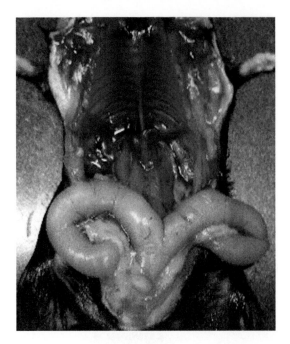

Figure 1.38   Bilateral mucometra in the female mouse uterus.

enlargement and convolution of the uterine horns is common and usually due to endometrial hyperplasia. The vagina is the short, tubular organ that leads from the cervix to the exterior. It is useful to remove the ovaries, uterus, urinary bladder and vagina together. The prosector should separate the ovaries from their attachments to the abdominal wall and pull the ovaries and uterine horns back to the level of the vagina and make an incision through the vaginal tube. Occasionally the vagina may be removed separately incorporating the external vulva.

The male reproductive organs consist of paired testes, epididymis, paired seminal vesicles and coagulating glands, prostate and penis. The mouse testes can move between the scrotum and the abdominal cavity. The male reproductive organs may all be removed together or they may be removed separately. The preputial glands are situated between the penis and rectum in the male animals and, in the female mouse, the clitoral glands are found in the same region, cranial to the vulva (Figure 1.39). Common macroscopic lesions noted at necropsy in the mouse

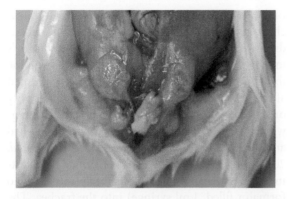

Figure 1.39   The preputial glands are situated between the penis and rectum in the male animals and in the female mouse, the clitoral glands are found in the same region, cranial to the vulva.

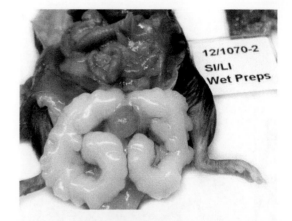

Figure 1.41   The seminal vesicles are situated on either side of the bladder and may be distended in older animals.

include bite wounds to the penis with resultant balanoposthitis (infection of the penis and prepuce), haemorrhage into the bulbourethral glands and often urinary bladder distension. This is referred to as mouse urologic syndrome and this condition may be increased in AKR mice housed in wire cages (Everitt *et al.* 1988). Preputial gland abscess is a common finding at necropsy in the older mouse (Figure 1.16). The testes are light tan in colour and may be pulled out of the scrotal sac for examination (Figure 1.40). Atrophy of one testis is common in male mice. The seminal vesicles are situated on either side of the urinary bladder and communicate with the urethra. They are generally white in colour, although older

mice may show tan-yellow discoloration as well as distension (Figure 1.41) (Finch and Girgis 1974). The testes should be removed with the epididymis intact. The seminal vesicles should be removed with an incision as low as possible to avoid the loss of seminal vesicular fluid. The urinary bladder and the prostate may be removed together and the penis and associated skin structures should also be removed separately. Fixation of the testes is complex and modified Davidson's fixative is recommended over Bouin's fixative (Lanning *et al.* 2002; Latendresse *et al.* 2002) or formalin. Modified Davidson's fixative is recommended because the morphological detail is good and there is minimal shrinkage of central tubules in the testes.

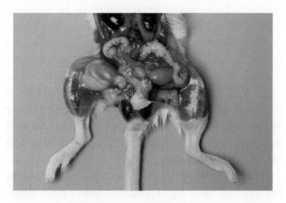

Figure 1.40   The testes are light tan in colour and may be pulled out of the scrotal sac for examination.

1.8   Removing the ribcage to expose lungs and heart

The prosector should now open the thoracic cavity by lifting the sternum and making a small incision in the diaphragm and then cutting through the ribs on both sides at the point where the ribs attach to the costal cartilage (Figure 1.42). The hyoid bones at the base of the larynx should now be incised. The thoracic cavity should be carefully examined for the presence of fluid (hydrothorax) or pus (pyothorax)

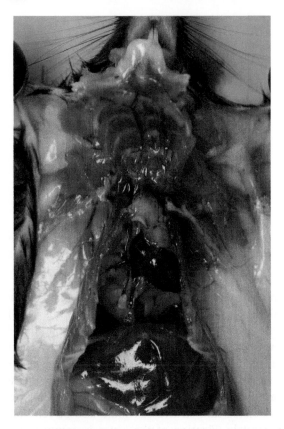

Figure 1.42   The prosector should now open the thoracic cavity by lifting the sternum (bone to which ribs are attached) and making a small incision in the diaphragm and then cutting through the ribs on both sides at the point where the ribs attach to the costal cartilage.

as well as adhesions between the lung lobes and the pleural cavity. Diaphragmatic hernias can result in strangulated liver lobes being present in the thoracic cavity.

Lung tissue consists of 90% air and 10% lung tissue (Braber *et al.* 2010). If the lung is placed, unperfused, into fixative, then collapse, deflation and disruption of the lung tissue occurs resulting in a histopathological section which resembles interstitial pneumonia and gives the impression of alveolar wall thickening and hypercellularity (van Kuppeveldt *et al.* 2000). For this reason, intratracheal instillation of 10% NBF (Braber *et al.* 2010) is recommended for optimal lung histopathological detail.

Braber *et al.* (2010) have demonstrated that intratracheal instillation of 10% NBF and paraffin embedding is superior to plastic embedding and Carnoy's instillation via tracheal instillation or fixed volume fixation. The tracheal instillation may be performed after removing the lungs from the thoracic cavity or *in situ* (which prevents the collapse of the lungs which occurs when the thoracic cavity is opened and the negative pressure is removed) (Braber *et al.* 2010). In general, perfusion is performed by inserting a plastic cannula or 22G needle (attached to formalin-filled, 1 ml syringe) into the trachea. The trachea may be pinched closed around the needle or cannula or a ligature may be tied around the trachea to keep the cannula in position. The syringe plunger is then depressed to inflate the lungs with formalin. The lungs are adequately inflated when the formalin reaches the margins of all lung lobes and the lungs fill the chest cavity. The prosector should see the lungs increase in size (Figure 1.43). The trachea is then grasped with forceps to keep the lumen closed while the lungs are immersed in formalin or if necessary the ligature is tied off.

Removal, of the tongue, trachea, oesophagus, lungs and heart as a unit is recommended (Seymour *et al.* 2004; Knoblaugh *et al.* 2011). This is performed by cutting through the muscles between the mandibles, pulling the tongue through to the outside of the animal and using forceps to gently remove the entire tongue, larynx, pharynx, trachea, oesophagus, heart, thymus and lungs from the neck area and the thoracic cavity. The heart and thymus may now be removed separately from the removed tissues, by cutting the heart at its base. The thymus may be extremely small in immunocompromised mice (such as severe combined immunodeficiency (SCID) mice) or due to atrophy in older mice making it difficult to locate. In this case tissue in the expected region of the thymus at the entrance to the thoracic cavity should be harvested and kept in a separate pot of formalin or cassette so that it can be identified later.

The heart is conical in shape and is divided by the septum into right and left areas each made up of an atrium and ventricle. Occasionally, the prosector may wish to bisect the heart at necropsy

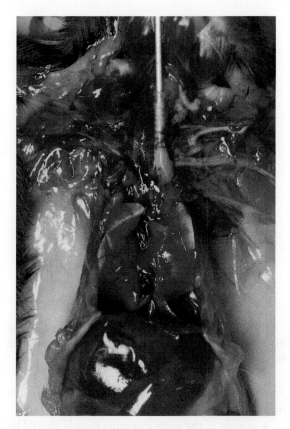

Figure 1.43   During the lung inflation procedure, the prosector should see the size of the lungs increase.

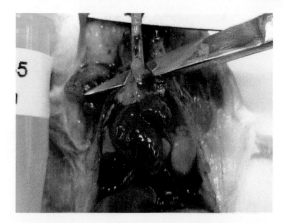

Figure 1.44   The mouse heart is conical in shape and in this animal there is a thrombus in the left atrium.

but generally the heart is fixed whole in order to avoid damage to the endocardium and valves. The pericardium should be examined for purulent exudate and fluid and the heart should be examined while attached to the blood vessels and lungs to look for congenital defects. Tumours are occasionally observed in the myocardium. Enlargement of the heart may be observed in cardiomyopathy, which is common in older mice. Cardiomyopathy is a diagnostic term used to describe a spectrum of spontaneous, age-related, degenerative changes, including degeneration, necrosis and increased interstitial fibrous tissue. The inflammatory component of these changes varies (Elwell and Mahler 1999). Left auricular thrombosis (blood clot in the heart chamber) may occur and is reportedly more common in older, breeding female mice with an

incidence of up to 66% (Meier and Hoag 1961). The thrombosis is visible at necropsy as a yellow mass within an enlarged left atrial chamber of the heart (Figure 1.44). Occasionally, the pericardium may be filled with blood due to the rupture of the atrium because of the presence of a large thrombus. BALB/c mice show epicardial mineralization (heart calcification) which increases with age with an incidence of 11% in males and 4% in female mice (Frith *et al.* 1975). The lesion may be recognized at necropsy by the presence of white, hard deposits on the epicardial surface of the heart (Figure 1.45).

The lungs are made up of four right lobes and one left lobe. The lungs should be spongy and

Figure 1.45   White cardiac mineralization of the epicardium of the mouse heart.

Figure 1.46   Lung adenoma is a common benign tumour of the mouse lung.

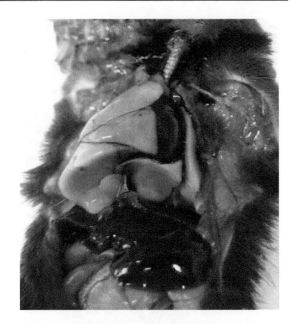

Figure 1.47   The lungs do not collapse upon opening the thoracic cavity in acidophilic macrophage pneumonia in the mouse.

elastic in consistency. The lobes should be examined carefully for the presence of atelectasis (collapse), oedema (proteinaceous fluid within the alveoli) and pneumonia (inflammation of the lung causing a firm consistency). Forty-four percent of Swiss mice more than six months old have lung tumours (Lynch 1969). These are generally benign tumours called adenomas and are made up of the alveolar walls of the lung (Figure 1.46).

Acidophilic macrophage pneumonia is an unusual form of disease within the lungs of the black C57BL/6 mice, which results in a cellular reaction around distinctive, red crystals. Ten per cent of C57BL/6 mice are susceptible (Murray and Luz 1990). At necropsy, the acidophilic pneumonia lungs are very prominent and do not collapse upon opening of the thoracic cavity (Figure 1.47). *Pneumocystis carinii* infection is common in immunocompromised mouse strains and may result in macroscopic lesions consisting of a rubbery consistency of the lungs, a failure of the lungs to collapse and the presence of multifocal, white areas in the lungs (Treuting *et al.* 2012).

The thyroids are located on either side of the lateral and dorsal surfaces of the trachea. If the salivary glands are removed, the thyroids are visible; however each thyroid is only 2–3 mm in length and they are generally not dissected out but are left attached to the trachea. Occasionally tumours may be noted within the thyroids. The parathyroids are located within the thyroids but are not visible macroscopically (Fiette and Slaoui 2011).

A sample of the gastrocnemius muscle should be taken from the lower hind limb. The femur should be removed and bone marrow may be collected from the femur for bone marrow cytological examination (Reagan et al., 2011). In addition, the entire femoral tibial joint should be removed and fixed for examination of the joint if required.

1.9   Removing the brain and spinal cord

The head should be separated from the body by flexing the head and cutting through the muscles and spinal cord at the level of the first cervical vertebra. The prosector should then remove the eyes from the head and should ensure that a section of optic nerve is included with the eye (Fiette and Slaoui 2011). The Harderian glands should be visible in the orbital cavity behind the eyeball and are usually

Figure 1.48 The head is separated from the body before removal of the brain. The Harderian glands should be visible in the orbital cavity behind the eyeball and are usually a pinkgrey colour.

Figure 1.49 Ocular discharge in the left eye and nasal barbering.

a grey to pink colour (Figure 1.48). The eyes can be fixed in formalin, but improved morphology may be obtained with Davidson's fixative (Chapter 12). Benign and malignant tumours of the Harderian glands are common in CD-1 mice (Maita *et al.* 1988) and may cause keratitis and prolapse of the eye. Some strains and stocks of mice have a high incidence of corneal opacities (DBA/2 (29.1%), C3H (16.2%), CF1 (16.2%) and BALB/c (10.0%)) whereas others have a lower incidence (CD-1 (4.3%) and C57BL/6 (4.1%)) (van Winkle and Balk 1986). In addition, severe inflammation of the eye – inflammation of the cornea and conjunctiva – may be caused by *Staphylococcus aureus* infection (Figure 1.49). Zymbal's glands are modified sebaceous glands situated at the base of the external ear (Fiette and Slaoui 2011) which secrete into the auditory canal. A sample of the sciatic nerve should be taken after locating the nerve within the biceps femoris muscle (Fiette and Slaoui 2011).

Small, pointed scissors should be inserted into the opening at the back of the brain (foramen magnum) and two incisions made through the occipital bone and then around the edge of the brain on either side, above the opening of the ear. The incisions are extended towards the nose and a cut is made across the frontal bone to form a flap, which may be lifted to visualize the brain. The brain can be gently removed by cutting the nerves and vessels at the base of the

brain, turning the skull upside down (Seymour *et al.* 2004) and allowing the brain to fall into a fixative container placed immediately below. Artefacts consisting of muscle and bone fragments inserted into the brain occur at this stage, thus caution is required. For some techniques optimal fixing of the brain requires perfusion (Knoblaugh *et al.* 2011), but this is not necessary for standard morphological assessment and the decision to perfuse must take into consideration the benefits to be gained versus the likelihood of inducing additional artefacts (see chapter 8). The brain should be examined *in situ* for the presence of inflammation (purulent material), haemorrhage, the presence of tumours and hydrocephalus (enlargement of the ventricles with thinning of the cerebral cortex). Once the brain has been removed it is important to visualize the pituitary gland (Figure 1.50). The pituitary gland remains in the sella turcica of the skull after the brain has been removed (Hagan *et al.* 2011). The gland is extremely friable and it may be advisable to section the bone on either side of the pituitary gland and to remove the pituitary gland with the bone attached and to fix the whole structure.

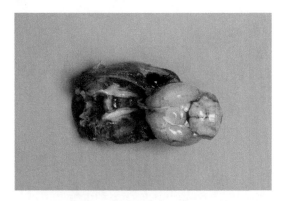

Figure 1.50   Once the brain has been removed it is important to visualize the pituitary gland.

1.10   Collecting and fixing tissue samples

In general, the histological examination of some or all tissues is necessary for the final diagnosis and thus it is important to know how to deal with taken during the necropsy. Slaoui and Fiette (2011) have reviewed the histopathology procedures from tissue sampling reviewed to histopathological evaluation. If the prosector is unable to find any gross abnormalities in a mouse that died suddenly, then it is advisable to collect small and large intestine, stomach, salivary glands, kidney, spleen, heart, lung, brain and liver into 10% neutral buffered formalin. These tissues will give the pathologist a reasonable idea about what might have happened in the animal.

All tissues should be fixed as soon as possible. Recommended fixatives include 10% neutral buffered formalin or 4% paraformaldehyde (Cardiff *et al.* 2000) or modified Davidson's solution for the fixing of testes and eyes. Approximately 20 times the volume of fixative should be used to the volume of the tissues to obtain optimal fixation (Seymour *et al.* 2004). Retaining the remainder of the carcass in formalin is good practice and will allow further analysis if necessary, for example examination of the nasal cavity if rhinitis is suspected or examination of tympanic bullae for otitis media.

Trimming is the process whereby fixed tissues collected during the necropsy are further dissected in order to fit into the embedding cassettes (Knoblaugh *et al.* 2011). If the tissue is too big for the cassette, then the cassette lid will cause impression marks on the tissue surface (Knoblaugh *et al.* 2011). Fresh tissue samples may also be frozen in OCT compound (Tissue Tek, UK) for immunohistochemistry or in situ hybridization. Excellent trimming and blocking patterns indicating how to section each tissue and which tissues should be placed together in a cassette are contained in Ruehl-Fehlert *et al.* (2003), Kittel *et al.* (2004), and Morawietz *et al.* (2004).

References

Braber, S., Verheijden, K.A., Henricks, P.A. *et al.* (2010) A comparison of fixation methods on lung morphology in a murine model of emphysema. *American Journal of Physiology: Lung Cellular and Molecular Physiology* **299**, L843–851.

Carbone, L., Carbone, E.T., Yi, E.M. *et al.* (2012) Assessing cervical dislocation as a humane euthanasia method in mice. *Journal of the American Association for Laboratory Animal Science* **51**, 352–356.

Cardiff, R.D., Anver, M.R., Gusterson, B.A. (2000) The mammary pathology of genetically engineered mice: the consensus report and recommendations from the Annapolis meeting. *Oncogene* **19**, 968–988.

Clarke, M.C., Taylor, R.J., Hall, G.A. and Jones, P.W. (1978) The occurrence in mice of facial and mandibular abscesses associated with Staphylococcus aureus. *Lab Animal* **12**, 121–123.

Elwell, M.R. and Mahler, J.F. (1999) Heart, blood and lymphatic vessels, in *Pathology of the Mouse, Reference and Atlas* (eds R.R. Maronpot, G.A. Boorman and B.W. Gaul), Cache River Press, Vienna, pp. 361–380.

Everitt, J.I., Ross, P.W. and Davis, T.W. (1988) Urologic syndrome associated with wire caging in AKR mice. *Laboratory Animal Science* **38**, 609–611.

Fiette, L. and Slaoui, M. (2011) Necropsy and sampling procedures in rodents. *Methods in Molecular Biology* **691**, 39–67.

Finch, C.E. and Girgis, F.G. (1974) Enlarged seminal vesicles of senescent C57BL-6J mice. *Journal of Gerontology* **29**, 134–138.

Frankenberg, L. (1979) Cardiac puncture in the mouse through the anterior thoracic aperture. *Lab Animal* **13**, 311–312.

Frith, C.H., Haley, T.J. and Seymore, B.W. (1975) Spontaneous epicardial mineralization in BALB/cStCrl mice. *Laboratory Animal Science* **25**, 787.

Frith, C.H. and Ward, J.M. (1988) *Color Atlas of Neoplastic and Non-Neoplastic Lesions in Aging Mice*, Elsevier, Amsterdam.

Goto, N., Nakajima, Y., Onodera, T., *et al.* (1984) Inheritance of hydronephrosis in the inbred mouse strain DDD. *Lab Animal* **18**, 22–25.

Hagan, C.E, Bolon, B. and Keene, C.D. (2011) Nervous system, in *Comparative Anatomy and Histology: A Mouse and Human Atlas* (eds P.M. Treuting and S. Dintzis), Elsevier, Amsterdam, pp. 339–394.

Hayat, M.A. (2000) Chemical fixation, in *Principles and Techniques of Electron Microscopy. Biological Applications* (ed. M.A. Hayat), 4th edn. Cambridge University Press, Cambridge, pp. 63–73.

Hoff, J. (2000) Methods of blood collection in the mouse. *Lab Animal* **29**, 47–53.

Kittel, B., Ruehl-Fehlert, C., Morawietz, G. *et al.* (2004) Revised guides for organ sampling and trimming in rats and mice – Part 2. A joint publication of the RITA and NACAD groups. *Experimental and Toxicologic Pathology* **55**, 413–431.

Knoblaugh, S., Randolph Habecker, J. and Rath S. (2011) Necropsy and histology, in *Comparative Anatomy and Histology: A Mouse and Human Atlas* (eds P.M. Treuting and S. Dintzis), Elsevier, Amsterdam, pp. 15–41.

Lanning, L.L., Creasy, D.M., Chapin, R.E. *et al.* (2002) Recommended approaches for the evaluation of testicular and epididymal toxicity. *Toxicologic Pathology* **30**, 507–520.

Latendresse, J.R., Warbrittion, A.R., Jonassen, H. and Creasy, D.M. (2002) Fixation of testes and eyes using a modified Davidson's fluid: comparison with Bouin's fluid and conventional Davidson's fluid. *Toxicologic Pathology* **30**, 524–533.

Lewis, D.J. (1984) Spontaneous lesions of the mouse biliary tract. *Journal of Comparative Pathology* **94**, 263–271.

Lynch, C.J. (1969) The so-called Swiss mouse. *Laboratory Animal Care* **19**, 214–220.

Maita, K, Hirano, M., Harada, T. *et al.* (1988) Mortality, major cause of moribundity, and spontaneous tumors in CD-1 mice. *Toxicologic Pathology* **16**, 340–349.

Margolis F.L. (1971) Regulation of porphyrin biosynthesis in the Harderian gland of inbred mouse strains. *Archives of Biochemistry and Biophyses* **145**, 373–381.

McInnes, E.F., Rasmussen, L, Fung, P. *et al.* (2011) Prevalence of viral, bacterial and parasitological diseases in rats and mice used in research environments in Australasia over a 5-y period. *Lab Animal (NY)* **40**, 341–350.

Meier, H. and Hoag, W.G. (1961) Studies on left auricular thrombosis in mice. *Experimental Medicine and Surgery* **19**, 317–322.

Michael, B., Yano, B., Sellers, R.S. *et al.* (2007) Evaluation of organ weights for rodent and non-rodent toxicity studies: a review of regulatory guidelines and a survey of current practices. *Toxicologic Pathology* **35**, 742–750.

Moolenbeek, C. and Ruitenberg, E.J. (1981) The 'Swiss roll': a simple technique for histological studies of the rodent intestine. *Lab Animal* **15**, 57–59.

Morawietz, G., Ruehl-Fehlert, C. and Kittel, B. *et al.* (2004) Revised guides for organ sampling and trimming in rats and mice – Part 3. A joint publication of the RITA and NACAD groups. *Experimental and Toxicologic Pathology* **55**, 433–449.

Murray, A.B. and Luz, A. (1990) Acidophilic macrophage pneumonia in laboratory mice. *Veterinary Pathology* **27**, 274–281.

Nakajima, K. and Ohi, G. (1977) Aerophagia induced by the nasal obstruction on experimental animals. *Jikken Dobutsu* **26**, 149–159.

Pearson, G.R. and Logan, E.F. (1978a) Scanning electron microscopy of early postmortem artefacts in the small intestine of a neonatal calf. *British Journal of Experimental Pathology* **59**, 499–503.

Pearson, G.R. and Logan, E.F. (1978b) The rate of development of postmortem artefact in the small intestine of neonatal calves. *British Journal of Experimental Pathology* **59**, 178–182.

Reagan, W.J., Irizarry-Rovira, A., Poitout-Belissent, F. *et al.* (2011) Bone Marrow Working Group of ASVCP/STP. Best practices for evaluation of bone marrow in nonclinical toxicity studies. *Toxicologic Pathology* **39**, 435–448.

Ruehl-Fehlert, C., Kittel, B., Morawietz, G. *et al.* (2003). Revised guides for organ sampling and trimming in rats and mice – Part 1. A joint publication of the RITA and NACAD groups. *Experimental and Toxicologic* **55**, 91–106.

Sarna, J.R., Dyck, R.H. and Whishaw, I.Q. (2000) The Dalila effect: C57BL6 mice barber whiskers by plucking. *Behavioural Brain Research* **108**, 39–45.

Sellers, R.S., Morton, D., Michael, B. *et al.* (2007) Society of Toxicologic Pathology position paper: organ weight recommendations for toxicology studies. *Toxicologic Pathology* **35**, 751–755.

Seymour, R., Ichiki, T., Mikaelian, I. *et al.*. (2004) Necropsy methods, in *The Laboratory Mouse* (eds H. Hedrich and G. Bullock), Elsevier, New York, pp. 495–517.

Sheldon, W.G., Greenman, D.L. (1980) Spontaneous lesions in control BALB/C female mice. *Journal of Environmental Pathology and Toxicology* **3**, 155–167.

Shirama, K., Furuya, T., Takeo, Y. *et al.* (1981) Influences of some endocrine glands and of hormone replacement on the porphyrins of the Harderian glands of mice. *Journal of Endocrinology* **91**, 305–311.

Slaoui, M. and Fiette, L. (2011) Histopathology procedures: from tissue sampling to histopathological evaluation. *Methods in Molecular Biology* **691**, 69–82.

Slauson, D.O. and Cooper, B.J. (2000) *Hypoxic cell injury, in* Mechanisms of Disease. A Textbook of Comparative General Pathology, Mosby, St Louis, pp. 45–53.

Smith, R.S., Roderick, T.H. and Sundberg, J.P. (1994) Microphthalmia and associated abnormalities in inbred black mice. *Laboratory Animal Science* **44**, 551–560.

Son, W.C. (2003) Factors contributory to early death of young CD-1 mice in carcinogenicity studies. *Toxicology Letters* **145**, 88–98.

Sundberg, J.P., Elson, C.O., Bedigian, H. and Birkenmeier, E.H. (1994) Spontaneous, heritable colitis in a new substrain of C3H/HeJ mice. *Gastroenterology* **107**, 1726–1735.

Sundberg, J.P., King, L. (1996) Cutaneous changes in commonly used inbred mouse strains and mutant stocks, in *Pathobiology of the Aging Mouse* (eds U. Mohr, J.M. Ward, C.C. Capen *et al.*), ISlI Press, Washington, DC, pp. 325–337.

Suttie, A.W. (2006) Histopathology of the spleen. *Toxicologic Pathology* **34**, 466–503.

Suwa, T., Nyska, A., Peckham, J.C. *et al.* (2001) Maronpot RR.A retrospective analysis of background lesions and tissue accountability for male accessory sex organs in Fischer 344 rats. *Toxicologic Pathology* **29**, 467–478.

Taradach, C. and Greaves, P. (1984) Spontaneous eye lesions in laboratory animals: incidence in relation to age. *Critical Reviews in Toxicology* **12**, 121–147.

Treuting, P.M., Clifford, C.B., Sellers, R.S. and Brayton, C.F. (2012) Of mice and microflora: considerations for genetically engineered mice. *Veterinary Pathology* **49**, 44–63.

Valentine, H., Williams, W.O. and Maurer, K.J. (2012) Sedation or inhalant anesthesia before euthanasia with CO_2 does not reduce behavioral or physiologic signs of pain and stress in mice. *Journal of the American Association for Laboratory Animal Science* **51**, 50–57.

Van den Broeck, W., Derore, A. and Simoens, P. (2006) Anatomy and nomenclature of murine lymph nodes: descriptive study and nomenclatory standardization in BALB/cAnNCrl mice. *Journal of Immunological Methods* **312**, 12–19.

van Kuppevelt, T.H., Robbesom, A.A., Versteeg, E.M. *et al.* (2000) Restoration by vacuum inflation of original alveolar dimensions in small human lung specimens. *European Respiratory Journal* **15**, 771–777.

Van Winkle, T.J. and Balk, M.W. (1986) Spontaneous corneal opacities in laboratory mice. *Laboratory Animal Science* **36**, 248–255.

Zurcher, C.M., van Zwieten, M.J., Solleveld, H.A. and Hollander, C.F. (1982) Aging research, in *The Mouse in Biomedical Research: Experimental Biology and Oncology* (eds H.L Foster, J.D. Small and J.G. Fox), Academic Press, Amsterdam, pp.11–35.

Chapter 2

Practical approaches to reviewing and recording pathology data

Cheryl L. Scudamore

Mary Lyon Centre, MRC Harwell, UK

Pathology is often thought of as a descriptive subject generating observational data in the form of a narrative report (Holland 2011). This is based on the approach that is commonly used in diagnostic pathology where a pathology report comprises a paragraph describing the morphological features of any findings present usually for a single or limited number of tissues from one patient. The report will conclude with a diagnosis and often some prognostic interpretation. This is the commonly used approach familiar to most conventionally trained veterinary and medical diagnostic pathologists reporting on necropsy (macroscopic post mortem observations and follow up microscopic samples) or biopsy (microscopic preparations) material. In experimental pathology, the scientist or pathologist will usually be evaluating a large number of tissues from many animals and will be looking for differences in the incidence and severity of any changes present. A descriptive approach to pathology reporting can be used in this situation but it leads to the generation of qualitative and ordinal data, which can be difficult to manipulate and analyse statistically (Table 2.1). In addition, if consistent lesion terminology is not rigorously applied it can be very difficult to compare individual or groups of animals. Narrative descriptions are also time consuming to produce and can significantly slow down the reporting of experimental results and so more simplified and efficient approaches are needed for experimental pathology.

Two approaches can be used to help generate data more quickly from pathology specimens and in a form that can be more easily statistically manipulated. The first technique is to use standardized terminology for the description of lesions and the second is to apply some form of

A Practical Guide to the Histology of the Mouse, First Edition. Cheryl L. Scudamore.
© 2014 John Wiley & Sons, Ltd. Illustrations © Veterinary Path Illustrations, unless stated otherwise.
Published 2014 by John Wiley & Sons, Ltd. Companion Website: www.wiley.com/go/scudamore/mousehistology

Table 2.1 Comparison of diagnostic and experimental approaches to reporting pathology data adapted from Holland 2011.

Diagnostic approach	Experimental approach
Expecting to find pathology to support a 'diagnosis'	No expectation of identifying specific findings. Testing an hypothesis
Pathologist 'compares' findings with historical memory	In addition pathologist compares findings with concurrent controls
Uses knowledge of similar lesions in same species/strain or different species (e.g. from the literature and previous experience)	Also uses knowledge from concurrent controls and historical controls from similar strain
May produce a definitive, differential diagnosis or a diagnosis based on exclusion	Result is a list of difference between controls and treatment group animals
Rarely peer reviewed, but may request second opinion	Often peer reviewed (toxicological pathology)
Qualitative narrative report	Semiquantitative data presented in form of incidence and severity tables
No statistical analysis required or feasible	Statistical analysis is feasible and may be applied/necessary

scoring or quantitative methodology to enrich the data. These approaches also help to ensure the quality of pathology data, which can be evaluated using three parameters: thoroughness, accuracy and consistency (Shackleford *et al.* 2002). 'Thoroughness' relates to the recording of all observations including background lesions and requires that the pathologist or scientist is aware of the range of normal anatomical and histological appearances of tissues. Observations or 'findings' can include anatomical variations, such as ectopic splenic nodules, physiological changes, such as those seen in the reproductive tract during the oestrus cycle, or background lesions, which are nonanatomical changes commonly seen in a given species, strain or age of animals (McInnes 2011a). Correctly recognizing changes equates to 'accuracy' and requires a familiarity with, and understanding of, the pathological changes that may occur, which can be aided by reference to the available published glossaries and nomenclatures. Consistency refers both to the correct use of terms for any given finding and the application of comparable terms for severity each time – that is, the use of a grading or quantitation system if required. The usefulness of these techniques are also greatly assisted by consistent sample selection and tissue trimming patterns.

2.1 Sample selection and trimming patterns

In any experimental system there is a necessity for controls and defined protocols to ensure repeatable and valid results. Histology is no different from any other laboratory technique in this respect and standardization of tissue sampling, trimming and sectioning greatly improves the quality and reproducibility of data that can be obtained from tissue sections. Consistent presentation of tissue sections on slides also reduces the amount of time taken to read the slides, reduces pathologist fatigue and reduces the risk of lesions being missed. For any given study a subset of tissues of interest will usually be identified in advance and harvested systematically at necropsy (Chapter 1). Once adequately fixed, tissue samples must then be selected, trimmed and embedded prior to sectioning. Standardization of the site of tissue to be sampled, amount of tissue to be evaluated and orientation of the tissue to be sectioned all help to improve consistency. The specific requirements of the study will determine what tissue type is examined but in general the probability of observing a specific lesions relates to how much tissue is examined. In practice the amount of tissue that is examined is inevitably constrained by the costs of histology and pathologist time and so, in practice, it is useful to have standardized techniques that maximize the chances of identifying lesions while minimizing the cost. Published guides for sampling and trimming most mouse tissues are available (Ruehl-Fehlert *et al.* 2003; Kittel *et al.* 2004; Morawietz *et al.* 2004) but it may be useful to produce in-house guides that detail the requirements of a specific study (Figure 2.1).

A tissue trimming guide should include:

- tissue to be sampled;
- site of sampling – may specify region, such as kidney cortex or liver lobe;
- number of samples – for example, single section, multiple sections from different lobes/areas;
- plane of sampling – cross section or longitudinal;

- size of sample;
- orientation of tissue in block – for example, which tissue or region is closest to slide label or which face of the tissue (up or down) is sectioned of the tissue;
- whether left or right samples should be identified separately – for example, unilateral lesions;
- separation of tissues that look the same when fixed/processed – for example, identification of the individual lymph nodes.

There is often a temptation to try to reduce costs by presenting as many tissues as possible in a single wax block. This is often a false economy as tissues may be lost or poorly orientated and the blocks may be more difficult for histologists to produce. The numbers of wax blocks that are needed can be reduced by grouping tissues of similar density into

Guide to identification of lung lobes at necropsy

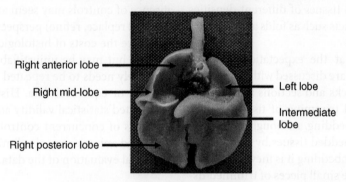

Right anterior lobe

Right mid-lobe

Right posterior lobe

Left lobe

Intermediate lobe

Guide for trimming and embedding lung lobes for histology

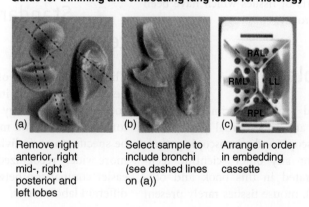

(a) Remove right anterior, right mid-, right posterior and left lobes

(b) Select sample to include bronchi (see dashed lines on (a))

(c) Arrange in order in embedding cassette

Figure 2.1 Example of a pictorial guide for identification of organ at necropsy and sampling of tissue for histology.

a single block – for example, liver, kidney, spleen and thymus (an example is given in Table 2.2). For complex or very small tissues, such as eyes, ovary, pituitary and adrenal glands, it may be difficult to orientate tissues to get representative sections of all the tissue components if they are blocked with other tissues, for example to achieve a good section of adrenal cortex and medulla it is best to place the adrenals in a separate cassette. Specific regions of intestine may be difficult to identify, particularly when modified by pathological changes and so the embedding schedule should be organized to aid identification. This can be achieved through a number of approaches, including placing the different sections of small or large intestinal regions that have histological similarities in separate blocks (see Table 2.2), using cassette dividers to separate tissues or ensuring that tissues are always orientated in a set pattern on the slide. Hard tissues which have been decalcified will also be better in a separate block as it is hard to section tissues of different densities without creating artefacts such as folds and wrinkles (McInnes 2011b).

It is important that the expectations for the histopathology results are discussed with the pathologist before tissue blocks and sections are made so that the selection and trimming of tissues can be optimized before embedding. Although it is possible to reorientate embedded tissues by melting the paraffin wax and re-embedding it is inevitably more difficult to re-orientate small pieces of trimmed tissue at this stage and impossible to change the sample or retrieve other parts of tissues which have been disposed of after necropsy.

2.2   Controls

Control animals and proper experimental design (Zeiss *et al.* 2012) are an essential part of the evaluation of macroscopic and microscopic pathology, as they are for any other scientific technique. As demonstrated in this book and others (McInnes 2011b), mouse tissues rarely present with the unaltered hypothetical histology seen in textbooks because the tissue is usually affected by the appearance of common background lesions, histological or anatomical variations and artefacts. Background lesions will vary with strain, age of mouse, laboratory, health status, environmental and nutritional conditions. Controls are therefore vital to allow comparison of these background changes (in the specific laboratory/animal house) and any induced changes in the mouse populations and to distinguish these from genuine treatment or experimentally induced effects including those induced by genetic modification. Although historical control data may be useful it cannot account for changes that may occur during the time course of an experiment in a particular laboratory. Examples of this type of variation include the potential effect of high fat diets on liver histology and clinical chemistry parameters and the effect of a subclinical infection in the colony on affected tissues. Reduction in numbers or absence of controls may seem attractive from a 3Rs (reduce, replace, refine) perspective (Wells 2011) or to reduce the costs of histological and pathological analysis but may be unjustifiable if an experiment ultimately needs to be repeated because of uninterpretable or equivocal data. Historical control data has limited statistical validity and absent or limited numbers of concurrent control animals compared to treated animals may also limit the validity of any statistical evaluation of the data.

2.3   Standardizing terminology

The use of glossaries of standardized terms can be helpful in generating incidence data for lesions encountered in pathology. Glossaries can be developed for macroscopic and microscopic findings and can be specific to an individual laboratory or part of a more widely recognized nomenclature allowing for easier comparison between analyses performed in different laboratories.

Table 2.2   Example of a blocking pattern for full phenotypic evaluation of mouse tissues. Blocks 1–15 give a good overview of tissues and 16–19 are advisable for completeness. LS = longitudinal section. TS = transverse section (Kittel *et al.* 2004; Moraweitz *et al.* 2004; Ruehl-Fehlert *et al.* 2004). More detailed guidance for specific investigations of individual organs is provided in the relevant chapters.

Block number	Tissues	Orientation/comments
1	Thymus, mesenteric lymph node, salivary glands	Thymus – representative LS sample Mesenteric lymph node – representative LS sample Salivary glands – LS to include all three major glands
2	Heart, skeletal muscle	Heart – LS section through all four chambers Muscle – TS of gastrocnemius
3	Liver, kidney, spleen	Liver – TS from left lateral and TS left and right medial lobes including gall bladder Kidney – TS from each organ or TS from left and LS from right (ensuring papilla included in both sections) Spleen – TS from largest part of organ
4	Lung	Lung – LS of four large lung lobes (remove accessory lobe) Inflate with fixative
5	Thyroid, parathyroid, aorta, oesophagus	Aorta – TS from thoracic aorta (brown fat often around aorta) Oesophagus – TS Thyroid, parathyroid – TS through thyroid and trachea Trachea – separate TS from below thyroid if needed
6	Adrenals	Adrenal glands – LS embed on any surface Ensure medulla is present
7	Brain	3 × TS sections Can include spinal cord
8	Ovaries or testes and epididymides	Ovary – if possible LS to include ovary, oviduct and tip of uterine horn Testes – TS from both or one TS and one LS Epididymides- LS from both
9	Urinary bladder, uterus and vagina or prostate and seminal vesicles	Urinary bladder – LS Uterus – TS both horns Vagina – LS through vagina, cervix and uterine body Seminal vesicle/coagulating gland – TS both sides Prostate – LS after careful positioning of dorsal and ventral lobes
10	Stomach, duodenum, jejunum, colon, pancreas	Stomach – LS sections through greater and lesser curvature including glandular and nonglandular Intestines – TS of each region Pancreas – LS left lobe
11	Ileum, caecum, rectum	Intestines – TS of each region
12	Skin, mammary	LS – from thoracic area to maximize chances of mammary tissue
13	Eyes (with optic nerve), Harderian gland	Eyes – orientate to include optic nerve Harderian gland and lens – LS and lens
14	Pituitary	Pituitary - LS
15	Sternum (includes bone marrow)	Sternum – LS to include marrow
16	Femur and stifle joint	Femur – LS to include stifle joint
17	Spinal cord	Spinal cord – following decalcification TS of cervical, thoracic and lumbar regions
18	Peripheral nerve (usually sciatic)	Nerve – LS
19	Brown and white fat	Brown fat – TS from dorsal shoulder fat pad White fat – LS perigenital fat pad

2.3.1 *Macroscopic terminology*

Observation of changes at necropsy may be the end point in a study if no further microscopic analysis is to be performed. In this situation, information gathered during the dissection may form the basis of incidence data from the experiment. However, in many instances the observations made at necropsy may not only provide a data set in themselves but may be useful pointer to/ or correlate with the microscopic data.

Unless observations at necropsy are the result of a known intervention, for example measuring the size of a xenograft tumour and counting metastatic nodules, no assumption should be made about the pathological basis of any changes observed. Lesions of similar appearance may have very different underlying pathology and aetiologies – for example an enlarged mouse spleen may be due to lymphoma, other neoplasm types, increased extramedullary haematopoiesis, lymphoid hyperplasia or congestion (Chapter 9). Other 'apparent' abnormalities may relate to physiological/agonal changes such as hypostatic congestion, post-mortem change, artefacts or even normal anatomical features such as Peyer's patches in the intestine. This may be confusing to the novice. It is therefore safer to describe what is seen during the necropsy rather than make a judgement about the pathology it may represent. A 'wrong' macroscopic diagnosis may be misleading if there is no histological follow up or may need to be explained in subsequent reports if microscopic follow up refutes the observational 'guess'. In summary, the use of diagnostic terms should be avoided and macroscopic changes recorded should be limited to descriptive terms.

The extent of data recording that is required should be considered before the time of necropsy and should be detailed in a protocol or standard operating procedure. It may be useful to have a tick sheet to ensure that all tissues are taken, weighed and described as necessary (Chapter 1). An agreed lexicon of terms should be followed, which should take into account the anatomical location of any lesions their size, shape, distribution and colour (King *et al.* 2006). The use of standardized terminology will help to reduce interobserver variability if multiple people are involved in necropsy of animals from a specific study or set of studies. To ensure systematic recording of lesions it may be useful to have an illustrated guide (Figure 2.1), particularly for the anatomy of major organs of interest. In addition it is useful to be aware of the terminology used to describe the axial location of lesions (Figure 2.2). It may be appropriate to record size based on measurements (in millimetres) or as a proportion of the organ or area affected. Terms such as enlarged or reduced in size should be avoided unless they are justified for example by a measurable change in organ dimensions or weight.

An example of a glossary of terms for macroscopic descriptions is given below (Table 2.3) with an illustration of distribution terms in Figure 2.3. If a number of different people are involved in recording necropsy images it may be useful to create an illustrated glossary with photographic examples of each term that is to be used to improve consistency.

At necropsy it may also be useful to keep a record of the lesion or supplement the written description with a photographic image. While it is relatively easy to acquire digital images, acquiring images which are good can be difficult (McGavin 2012). The organ or lesion of interest should be in the centre of the picture, occupying most of the frame, ideally with a measurement scale or ruler and photographed against a neutral nonreflective background. It is also important that unique identification data is captured with the image and the file is labelled so that it can be uniquely identified once stored electronically. Digital images may be inadmissible or invalid from a legal evidence or good laboratory practice (GLP) point of view and so care should be taken to ensure that the potential uses and usefulness of digital images is understood (Suvarna and Ansary 2001).

2.4 Microscopic terminology

Mouse tissues, in common with those of all other animals, show a range of background lesions,

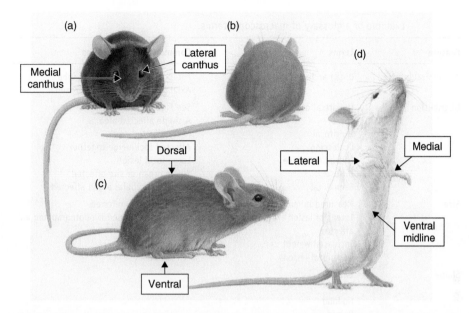

Figure 2.2   Anatomical land mark descriptors. (a) Cranial view of a mouse. (b) Caudal view of a mouse. (c) Lateral view of mouse. (d) Standing mouse.

variation and common spontaneous pathology (Taylor 2011). It is important that these background changes are recognized and understood so that they are not misinterpreted as experimental or treatment-related changes (Schofield *et al.* 2012) particularly when encountered for the first time by scientists. Systematic recording of these background changes is also important to build up background data on the normal extent of variation within a strain or colony and to ensure that mouse phenotypes that result in alterations in the nature or incidence or severity of these changes are recognized. Spontaneous findings change with the age and strain of mice so it is important that researchers are also aware of these variations (Pettan-Brewer and Treuting 2011; Taylor 2011; Brayton *et al.* 2012, Table 2.4). In general, young mice (<4 months old) used in phenotyping studies show very few background lesions (Table 2.4) other than small foci of inflammatory cells (usually predominantly mononuclear lineages) scattered within the parenchyma of various organs.

Nomenclatures/glossaries that list and describe common background changes in rodents are published mainly by the toxicological pathology community (for an example see Thoolen *et al.* 2010) but ontologies for anatomy (MA mouse anatomy) and pathological findings (MPATH mouse pathology) are available (Schofield *et al.* 2010). Many of the nomenclatures/glossaries have been devised by and are aimed at recording data from safety studies used in the development and registration of new chemical entities (including pharmaceuticals and agrochemicals). However, because they are comprehensive and well recognized by many experienced rodent pathologists, they form a useful guide for identifying and recording spontaneous lesions for the scientist or pathologist who lacks experience with mouse tissues. Lesions that are specific to the phenotype or experimental manipulation can therefore be recognized as out with the normal range of spontaneous lesions. On first recognition, these novel observations will need to be described and named with reference to the norms for the species or if the phenotype recapitulates a human disease the terminology appropriate for that disease. It is highly recommended that confirmation of all findings, especially those that appear novel,

Table 2.3   Example of a glossary of macroscopic terms.

Feature	Terms	Comments and examples
Anatomical location	Organ and subdivision	For example, liver, specific lobe
		Axial location, e.g. cranial/caudal
Distribution	Unilateral or bilateral	See Figure 2.3
	Focal	A single affected area
	Multifocal	Many affected areas
	Coalescing	Affected foci merge together
	Locally extensive	Large focal lesion
	Diffuse	Whole organ or site affected
	Segmental	Section of tubular organ affected
Size	Measured in mm or cm	Measurements preferred.
	Extent of lesion – percentage affected	Ensure scale is used if photographing an image
	Increased weight	
	Number of lesions	
Shape	Circular	
	Ovoid	
	Nodular	
	Raised	
	Laminated	
	Punctate	
	Depressed	
Colour		Keep to simple primary colours.
		Additional simple descriptors
	Red	Usually accumulation of blood
	White	Multiple possibilities connective tissue, exudates
	Black	Blood, melanin, other pigments
	Brown	Pigments
	Green	Bile or putrefactive change
	Yellow	Usually bilirubin accumulation – icterus
	Pale	Use in conjunction with a colour term
	Dark	Use in conjunction with a colour term
	Stippled/streaked/mottled	Use in conjunction with a colour term
Consistency/texture	Hard	Like bone
	Firm	Like tendon
	Soft	Like brain
	Fluctuant	Like filled urinary bladder
	Friable	
	Gas or fluid filled	
	Gritty	
Contents	Volume	Keep separate samples of hollow organ or body cavity fluids if required
	Clear	
	Opaque	
	White	
	Yellow	
	Red	
	Frothy	
	Viscous	
	Gas	

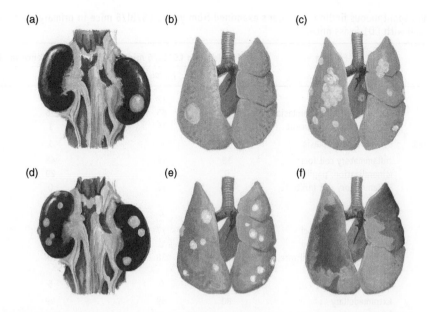

Figure 2.3   Examples of the use of terminology for describing macroscopic lesions. (a) and (b) could be described as (unilateral) focal lesions affecting one kidney or lung lobe respectively, (c) as multifocal to coalescing lesions, (d) as bilateral multifocal lesions, (e) as multifocal lesions affecting all lung lobes and (f) as locally extensive or diffuse lesions affecting left, right cranial and right middle lung lobes.

should be done in association with an experienced veterinary or medical pathologist as appropriate.

2.5   Recording pathology data

2.5.1   *Blinded versus unblinded review*

The decision as to whether to evaluate slides in a blinded or unblinded fashion is controversial (Dodd 1988 and Temple *et al.* 1988). In toxicological pathology, blinded reading of slides is considered unwise and the guidelines on best practice do not recommend it (Crissman *et al.* 2004; Neef *et al.* 2012). There are a number of reasons for this but, in general, for qualified pathologists, blinded review tends to go against their training, which, to avoid claims of negligent practice, requires them to integrate microscopic data with all other available sources of information (including, but not limited to, experimental protocol details, in life observations, macroscopic information, clinical pathology data, organ weight changes) to arrive at the best possible interpretation (Holland and Holland 2011a). Blinding may reduce the likelihood of identifying any changes, particularly those which represent a variation from 'normal' background findings (Neef *et al.* 2012). Evaluating tissue blind to treatment or group may not be unbiased as obvious tissue changes may effectively and rapidly 'unblind' the examination. Unless there is a justifiable reason, a blind review should not normally be recommended as it is unlikely to provide the best possible result. In effect a blind review may be biased if changes are obvious and if the tissues are being evaluated by an experienced pathologist the best use will not be made of their knowledge and interpretive skills.

A pragmatic solution that is commonly used in toxicological pathology (Holland and Holland 2011b) is to do an initial read of the tissues in an unblinded fashion to ensure that all findings are

Table 2.4 Common spontaneous findings in organs examined from young C57Bl/6 mice in primary phenotyping screens – comparison with CD1 Swiss mice.

Organ	Finding	Incidence in C57Bl/6 mice (%)		Incidence in CD1 mice (%)	
		Male	*Female*	*Male*	*Female*
Adrenal	X-zone vacuolation	0	12	0	53
	Subcapsular cell hyperplasia	13	65	16	60
	Corticomedullary pigment	0	15	1	6
Mesenteric lymph node	Lymphoid hyperplasia	13	0	1	0
Kidney	Inflammatory cell foci*	13	12	48	39
	Mineralization, papilla	0	4	29	20
	Inflammatory cell infiltrate, pelvis	13	12	0	0
	Basophilic tubules	0	4	32	24
Liver	Inflammatory cell foci	60	77	58	82
	Periportal vacuolation	20	0	12	12
	Inflammatory cell infiltrate, periportal	13	38	NR	NR
	Focal necrosis	7	0	5	6
Spleen	Extramedullary haematopoiesis	80	69	99	99
	Pigment	13	19	0	17
	Lymphoid hyperplasia	20	8	0	1
Testes	Tubular atrophy	7	NA	4	NA
Epididimydes	Inflammatory cell foci	7	NA	2.5	NA
Salivary gland	Inflammatory cell foci	13	4	24	39
Stomach	Junctional intestinal metaplasia	0	4	1	1
	Inflammatory cell infiltrate, submucosa	0	8	3	1
Brain	Mineralization, thalamus	7	0	0	0
Harderian Gland	Inflammation	NE	NE	40	36
Urinary bladder	Inflammatory cell foci	0	0	10	22

Notes: Scudamore unpublished data from C57Bl/6N or C57Bl/6J wild type mice less than 3 months of age from two phenotyping programmes. CD1 mouse data from contract research organization database animals less than 5 months of age. *a*: Inflammatory cell foci refers to small foci of mixed but generally mononuclear inflammatory cells found within organ parenchyma with no specific pattern of distribution. NA: not applicable. NE: tissue not examined. NR: not recorded .

detected. This also allows for findings to be seen in context of the whole animal (or the subset of tissues examined). An example of this would be that chronic changes in the heart may result in secondary pathological changes in other organs, for example congestion and oedema in the lungs. Following the initial unblinded read, organs that have apparent changes are re-reviewed in a blinded fashion to ensure that the change seen is distinguishable from

the range seen in controls and to score or rank findings between groups if appropriate.

2.5.2 Identifying differences between groups

In most experimental situations, the scientist or pathologist is attempting to identify whether there is a difference between tissues from control (ideally concurrent) animals or those that have been experimentally manipulated. Identifying obvious deviations from 'normal' histology may be relatively easy in a first-pass review of the tissues, however more subtle changes may be harder to detect with certainty. Where low numbers of animals are available, for example in early phenotyping pipelines, the level of certainty may remain low particularly if the possible change seen is a minor variation in the background 'pathology'. In this case there may be no option but to wait for more mice to become available to increase the certainty that the finding represents a real change from wild type. Where larger group sizes are available, there are various methods that can be used to facilitate detection of subtle changes (Holland and Holland 2011b). The methods may also produce data that is more or less suitable for statistical analysis.

Scoring methods are the ones most commonly used by toxicological pathologists (Holland 2001) because they are easy to understand and generally quick to perform. They are equally applicable to experimental pathology and phenotyping applications (Brayton *et al.* 2012). These methods require separation of lesions into scores or grades (ordinal classes of data) and form the basis of semiquantitative analysis (see next section). Grades may be numerical (0 to 5) or descriptive (minimal to severe) or based on the proportion of tissue affected. In theory, the more classes that are used the more sensitive the method, but in reality there is a tendency to use three to five classes as it becomes increasingly difficult to maintain consistency with larger numbers of classes. Data that are produced can be demonstrated via incidence and severity tables and can be analysed statistically, although there are limitations to the power of the method.

Ordering or 'ranking' all lesions from least affected to worst is in theory the most statistically sound method of evaluation. In this technique tissues are simply ranked from the least affected by a given lesion (or with no lesion present) to the most severely affected. Nonparametric statistical tests can then be applied to the ranked data (for an example of how this can be used see Ibanez *et al.* 2004). While in theory this sounds easy, it may be difficult

to distinguish (and remember) subtle variations between slides and large numbers of animals not showing the feature (nonresponders) will limit the effectiveness of statistical evaluation. In practice most pathologists do not use this technique because it is considerably more time consuming than the scoring method, particularly if there are large numbers of animals to evaluate.

Other methods that are less commonly used include identifying tissues which have findings outside an 'expected' range or comparing pairs of tissues from treated and control animals. Briefly, there are two methods for comparing tissues outside a specific range; the 'affected method' requires identification of findings outside an expected 'historical' normal range (this requires a high level of previous experience and familiarity with background lesions) or the 'outside-control' method which aims to identify findings outside the range of findings seen in the control group (which is generally easier for less experienced pathologists or scientists to achieve). Paired methods are sometimes used when checking for the consistent presence or absence of a lesion. In these methods pairs of tissues from experimental and control animals are compared. These methods are relatively insensitive and from a statistical analysis viewpoint are strongly affected by the presence of animals with no findings or lesions.

2.6 Quantitative versus semiquantitative analysis

In the experimental setting it is often necessary to decide whether to record data in a qualitative or quantitative context. When describing a dominant novel phenotype, a qualitative description of the lesions may be appropriate, but in most experimental settings some degree of quantitation is necessary. At its simplest this may mean the calculation of incidence data based on whether a finding is present or absent in a tissue, organ or animal. The necessity for more detailed quantitation should be considered carefully (Figure 2.4) as inevitably this requires more resources in terms of time or equipment but does

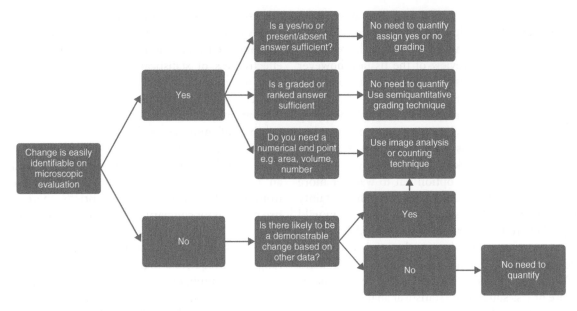

Figure 2.4   Quantification decision tree.

not systematically provide more significant data (for an example see Stidworthy *et al.* 2003). Physically counting individual cells, particles or tissue components on histological sections (profile-based counting) is costly in terms of pathologists' or scientists' time but, equally, image-analysis techniques may incur costs due to equipment and software outlay and time needed to programme/train the system to count the appropriate features. There are also a number of ways in which profile counts do not strictly correlate to the number of objects present in the counted sample and so, if required, the choice of method to use should be carefully considered ahead of the experimental study start (Coggeshall and Lekan 1996; Guillery and Herrup 1997; Schmitz and Hof 2005). It is therefore important that the case for quantitation is carefully considered and a cost benefit analysis performed in terms of the resources taken to obtain the data versus its eventual quality and usefulness.

The process of producing routine H&E stained tissue slides involves a number of steps including sample selection, fixation, tissue processing, embedding and orientation of sample, sectioning and staining, which can all lead to artefacts and degrees of variation in the appearance of the final section

(McInnes 2011b). This variation may lead to a large 'signal-to-noise' ratio which means that precise quantitation may be difficult and inappropriate. While immunohistochemistry (IHC) and *in situ* hybridization (ISH) techniques may appear to add more precision to the localization of specific cells, they are also prone to a range of process-related variation, which may equally reduce the value of exact quantitation (Moskaluk 2008). Variability in IHC staining can result from both pre-analytical and analytical sources (Dunstan *et al.* 2011). In addition to those factors involved in producing artefacts in the section, which are common with H&E staining, the time from sampling to fixation, type of fixative and fixation time can all significantly affect IHC outcomes (Scudamore *et al.* 2011). All of these factors need to be controlled as far as is possible, but it is important to realize the significant technical limitations to producing true quantitation in tissue sections prepared using any technique (Table 2.5).

The inevitable degree of variability in tissue section techniques and the degree of subjectivity in lesion assessment and interpretation means that semiquantitative techniques have a great deal of value and are routinely used in toxicological

Table 2.5   Variables in the process of producing an IHC stained section that can potentially affect quantitation.

Process	Variables
Tissue sampling	Time from surgery/ death to sampling
	Use of perfusion
Fixation	Type of fixative used
	Time in fixative
	Tissue thickness
Processing/embedding	Length of processing schedule
	Processing reagents
	Embedding matrix
	Orientation of tissue at embedding
Sectioning	Thickness of section
	Artefacts – cutting, spreading of section
IHC method	Antigen retrieval techniques – heat versus enzymatic
	Antibodies – source, dilutions
	Protocol – timing, temperatures
	Chromogen / detection method
	Blocking techniques
	Manual versus automated technique (stain variation)
Image analysis method	Defining area of interest
	Defining nonstaining area for comparison
	Identifying background staining
	Analysis of the intensity of staining or of the presence/absence of staining

pathology (Shackelford *et al.* 2002; Holland and Holland 2011a).

2.7   Semiquantitative techniques

Quantitative or 'parametric' tests are rarely used by toxicological pathologists who are used to dealing with high volumes of rodent pathology (Holland 2001). Comparisons between careful use of semi-quantitation and quantitative methods (manual or image-analysis based) suggest that there is often no significant effect on the overall conclusions that are reached (Shackelford *et al.* 2002; Von Bartheld 2002). Semiquantitative analysis is most commonly used for the recording of non-neoplastic pathological findings and is based on estimates of

the relative severity or extent of a lesion rather than precise measurements. Observations are divided into ordinal grades or scores, which may be based on the extent of affected tissue or cell types present. The number of scores or grades used is usually 3 to 5 plus a 0 score sometimes referred to as NAD (no abnormality detected) or WNL (within normal limits). These terms have subtly different inferences with NAD suggesting that there is no lesion at all in the tissue whereas WNL allowing for some minimal variation from normal within the tissue. The choice of numbers of scores or grades is up to the individual researcher but there appears to be an inverse relationship between the number of grades and the reproducibility of data (Shackelford 2002) as it becomes increasingly hard to define minor nuances between grades and increasingly impractical to remember the difference between them when looking at sections. These methods may appear subjective but, while minor differences in grade between observers (for example +/− one grade) are common and considered insignificant, larger variations in grading are uncommon between trained pathologists. Scores may be represented as +, ++, +++ or numerically (1, 2, 3, 4, 5) or by description minimal, slight, moderate, moderately severe, severe. Grades may also represent linear or nonlinear distributions of severity – example descriptors (modified from Shackelford 2002) are given in Table 2.6. The important point is that whatever system is used, it is used consistently between tissues or between studies in a given laboratory.

2.8 Quantitative techniques

In many situations researchers need to know if there is a change or difference in cellular features between groups but there is no added value or necessity in knowing the absolute numbers of cells that are present or stained with a particular technique. However, there are limits to the ability of the human eye to detect subtle changes in tissue, particularly with regard to cell number (Wanke 2002) and in situations where more detailed analysis is needed, for example examining subtle changes in neuroanatomical features, then quantitative techniques may be necessary (Boyce *et al.* 2010). A full discussion of the methods used for quantification and their advantages and disadvantages is beyond the scope of this chapter and is somewhat controversial; readers are referred to the referenced publications for more detailed information, however a brief outline is given below.

Quantitation can be achieved manually by an individual scientist counting 'events' by scanning a slide or an image of a slide or by computer-assisted techniques such as image analysis or laser scanning cytometry (Peterson *et al.* 2008). The analysis may be made on part of a slide or an image or on digitized data obtained from whole sections using whole slide imaging techniques (Dunstan *et al.* 2011).

Image analysis techniques may be useful for high contrast changes, for example when using IHC or special staining techniques that give a clear threshold between stained and unstained cells or tissue structures. A range of variables (Table 2.5) may affect the actual degree of staining with IHC, which may mean that although positive or negative cells or events may be counted there may be less confidence in relative estimates of antigen concentration. A number of issues arise when using image analysis, which increase the time and cost involved and reduce efficiency. It is usually necessary to determine the area of interest for analysis from the whole tissue – usually a region within the tissue. This takes time whether it is done manually or is automated following a period of programming or training the image analysis software to accurately identify the region. Thresholds also have to be set for the lower level of staining intensity which will be counted as positive staining above background. This step also usually requires manual intervention when setting up the technique.

There is much debate around the choice of counting techniques for estimating the total number of cells or particles in tissues (Coggeshall and Lekan 1996; Guillery and Herrup 1997; Von Bartheld 2002; Boyce *et al.* 2010) – determining the number of cells/volume of tissue. These methods rely on counting visible parts of cells or particles in a representative number of sections taken through the tissue to give an estimate of cells/volume. The traditional method of doing this is based on counting cell or particle 'profiles' in approximately every tenth section through the tissue of interest. This method counts the visible two dimensional appearance of a cell or part of a cell seen in a stained section and so are known as '2D', model-based or counting methods. An individual cell or particle may appear in more than one section depending on its size, orientation and shape and so, in order not to overestimate the numbers, assumptions have to be made about the cells, which then allow a correction factor to be applied to work out the estimate of cells/tissue volume (Abercrombie 1946). Despite the disadvantage of this technique in terms of biases that arise due to the need to make assumptions about particle size (further increased by small section thickness to particle height ratio or

Table 2.6 Examples of linear and nonlinear grading schemes that could be used for semiquantitative analysis of non-neoplastic lesions in mouse tissues.

Linear		Nonlinear	
Grade	*Description*	*Grade*	*Description*
0	NAD (WNL) – no change recorded	0	NAD (WNL) No change recorded
1	Minimal – 0–20% of tissue affected by change	1	Minimal – the least change that is visible on light microscopy at x20, small, focal or affect <10% of tissue
2	Mild (slight) – 21–40% of tissue affected by change	2	Mild (slight) – Change is readily detected but not a major feature, may be multifocal small lesions or affect <20% of tissue, may still be within background appearance for the species
3	Moderate – 41–60% of tissue affected by change	3	Moderate – Change is more extensive or is comprised of more foci (seen in every x20 field for example), beyond the usual background for the lesion in the species. May start to have relevance for organ function and may correlate with other changes for example increased organ weight.
4	Moderately severe (moderately marked) – 61–80% of tissue affected by change	4	Moderately severe (moderately marked) – as for 3 but more of the tissue is affected, e.g. up to 75%. Likely to have relevance for tissue/ organ function
5	Severe (marked) 81–100% of tissue affected by change	5	Severe (marked) – virtually the whole tissue is affected by the change and likely to be functionally relevant/detrimental

with irregular particle size) the relative ease of the technique and the possibility of estimating some of the errors allowing correction factors to be applied (Guillery and Herrup 1997) means that it is still overwhelmingly popular (Von Bartheld 2002).

Other quantitation techniques are known as 'design-based' or 3D techniques and these attempt to limit biases by counting events (the appearance of a cells/particles) in a real (optical dissector) or re-constructed (physical dissector) 3D space instead of profiles on a 2D section and by only counting those event in a fraction of the total region of interest chosen randomly so that objects have equal changes of being sampled independently of their size, shape or distribution. Practically this is done by comparing between two anatomically close sections (physical dissector) or using relatively thick tissue sections (optical dissector) or creating 3D reconstructions of tissue. The physical dissector and 3D recon-struction methods are rarely used because they are time consuming and require specialized equipment. The optical dissector is the most commonly used

design-based technique but is still rarely used com-pared to the 2D methods, (3D reported in 5% of papers using quantitation techniques versus 80% for 2D techniques – Von Bartheld 2002) and has been adopted in some fields, for example experimental neurology, more readily than others. In some cases particles may not be identifiable in thick sections and so this method is then ruled out.

The quantitation technique chosen used will also determine what histological techniques are used to prepare the sections. Paraffin embedded and cryosections are suitable for 2D and optical dis-sector techniques whereas thin resin sections are only suitable for 2D techniques and celloidin resin sections are better for the optical dissector approach.

Careful consideration should be made to take the decision to quantify cells or features in tissues particularly if this aims to establish absolute num-bers/tissue volume. Whatever method is chosen, pilot studies should be used to ensure the method is valid and calibrated for the required endpoint.

References

Abercrombie, M. (1946) Estimation of nuclear population from microtome sections. *Anatomical Record* **94**, 239–247.

Boyce, J.T., Boyce, R.W. and Gundersen, H.J. (2010) Choice of morphometric methods and consequences in a regulatory environment. *Toxicologic Pathology* **38**, 1128–1133.

Brayton, C.F., Treuting, P.M. and Ward, J.M. (2012) Pathobiology of aging mice and GEM: background strains and experimental design. *Veterinary Pathology* **49**, 85–105.

Coggeshall, R.E. and Lekan, H.A. (1996) Methods for determining numbers of cells and synapses: a case for more uniform standards of review. *Journal of Comparative Neurology* **364**, 6–15.

Crissman, J.W., Goodman, D.G., Hildebrandt, P.K. *et al.* (2004) Best practice guidelines: toxicologic histopathology. *Toxicologic Pathology* **32** (1), 126–131.

Dodd, D.C. (1988) Blind slide reading or the uninformed versus the informed pathologist. *Comment on Toxicology* **2**, 81–92.

Dunstan, R.W., Wharton, K.A. Jr., Quigley, C. and Lowe, A. (2011) The use of immunohistochemistry for biomarker assessment – can it compete with other technologies? *Toxicologic Pharmacology* **39** (6), 988–1002.

Guillery, R.W. and Herrup, K. (1997) Quantification without pontification: choosing a method for counting objects in sectioned tissues. *Journal of Comparative Neurology* **386** (1), 2–7.

Holland, T. (2001) A survey of discriminant methods used in toxicological histopathology. *Toxicologic Pathology* **29**, 269–273.

Holland, T (2011) Reporting of toxicologic histopathology: contrasting approaches in diagnostic versus experimental practice. *Toxicologic Pathology* **39**, 418–421.

Holland, T. and Holland, C. (2011a) Unbiased histological examination in toxicological experiments (or, the informed leading the blinded examination). *Toxicologic Pathology* **39**, 711–714.

Holland, T. and Holland, C. (2011b) Analysis of unbiased histopathology data from rodent toxicity studies (or, are these groups different enough to ascribe it to treatment?). *Toxicologic Pathology* **39**, 569–575.

Ibanez, C., Shields, S.A., El-Etr, M. *et al.* (2004) Systemic progesterone administration results in a partial reversal of the age-associated decline in CNS remyelination following toxin-induced demyelination in male rats. *Neuropathology and Applied Neurobiology* **30**, 80–89.

King, J.M., Dodd, D.C. and Roth, L. (2006) *The Necropsy Book*, Charles Louis Davis Foundation, Libertyville, IL.

Kittel, B, Ruehl-Fehlert, C., Morawietz, G. *et al.* (2004) Revised guides for organ sampling and trimming in rats and mice – Part 2. A joint publication of the RITA and NACAD groups. *Experimental Toxicologic Pharmacology* **55**, 413–431.

McGavin, M.D. (2012) Photographic techniques in veterinary pathology, in *Pathologic Basis of Veterinary Disease*, 5th edn (eds J.F. Zachary and M.D. McGavin), Elsevier, St Louis, MO, pp. 1245–1251.

McInnes, E.F. (2011a) Preface, in *Background Lesions in Laboratory Animals; A Colour Atlas* (ed. E.F. McInnes), Saunders, Edinburgh, p. vi.

McInnes, E.F. (2011b) Artifacts in histopathology, in *Background Lesions in Laboratory Animals; A Colour Atlas* (ed. E.F. McInnes), Saunders, Edinburgh, pp. 96–97.

Morawietz, G., Ruehl-Fehlert, C., Kittel, B. *et al.* (2004) Revised guides for organ sampling and trimming in rats and mice – Part 3. A joint publication of the RITA and NACAD groups. *Experimental Toxicologic Pharmacology* **55**, 433–439.

Moskaluk, C.A. (2008) The continued supremacy of qualitative interpretation over quantitative measurement in histologic tissue assessment. *American Journal of Clinical Pathology* **130**, 333–334.

Neef, N., Nikula, K., Francke-Carroll, S. and Boone, L. (2012) Regulatory forum opinion piece: blind reading of histopathology slides in general toxicology studies. *Toxicologic Pathology* **40** (4), 697–699.

Peterson, R.A., Krull, D.L. and Butler, L. (2008) Applications of laser scanning cytometry in immunohistochemistry and routine histopathology. *Toxicologic Pathology* **36** (1), 117–132.

Pettan-Brewer, C. and Treuting, P.M. (2011) Practical pathology of aging mice. *Pathobiology of Aging and Age-Related Diseases* **1**, 7202.

Ruehl-Fehlert, C., Kittel, B., Morawietz, G. *et al.* (2003) Revised guides for organ sampling and trimming in rats and mice – Part 1. A joing publication of the RITA and NACAD groups. *Experimental Toxicologic Pharmacology* **55**, 91–106.

Schmitz, C. and Hof, P.R. (2005) Design based stereology in neuroscience. *Neuroscience* **130**, 813–831.

Schofield, P.N., Gruenberger, M. and Sundberg, J.P. (2010) Pathbase and the MPATH ontology: community resources for mouse histopathology. *Veterinary Pathology* **47**, 1016–1020.

Schofield, P.N., Vogel, P., Gkoutos, G.V. and Sundberg, J.P. (2012) Exploring the elephant: histopathology in high-throughput phenotyping of mutant mice. *Disease Models and Mechanisms* **5**, 19–25.

Scudamore, C.L., Hodgson, H.K. *et al.* (2011) The effect of post-mortem delay on immunohistochemical labelling – a short review. *Comparative Clinical Pathology* **20**, 95–101.

Shackelford, C., Long, G., Wolf, J. *et al.* (2002) Qualitative and quantitative analysis of nonneoplastic lesions in toxicology studies. *Toxicologic Pathology* **30**, 93–96.

Stidworthy, M.F., Genoud, S., Suter, U. *et al.* (2003) Quantifying the early stages of remyelination following cuprizone-induced demyelination. *Brain Pathology* **13**, 329–339.

Suvarna, S.K. and Ansary, M.A. (2001) Histopathology and the 'third great lie'. When is an image not a scientifically authentic image? *Histopathology* **39**, 441–446.

Taylor, I. (2011) Mouse, in *Background Lesions in Laboratory Animals; A Colour Atlas* (ed. E.F. McInnes), Saunders, Edinburgh, pp. 45–72.

Temple R., Fairweather W. R., Glocklin V. C., O'Neill R. T. (1988). The case for blinded slide reading. *Comment on Toxicology* **2**, 99–109.

Thoolen, B., Maronpot, R.R., Harada, T. *et al.* (2010) Proliferative and nonproliferative lesions of the rat and mouse hepatobiliary system. *Toxicologic Pathology* **38** (7 Suppl.), 5S–81S.

Von Bartheld, C.S. (2002) Counting particles in tissue sections: choices of methods and importance of calibration to minimise biases. *Histology and Histopathology* **17**, 639–648.

Wanke, R. (2002) Stereology – benefits and pitfalls. *Experimental and Toxicologic Pathology* **54**, 163–164.

Wells, D.J. (2011) Animal welfare and the 3Rs in European biomedical research. *Annals of the New York Academy of Sciences* **1245**, 14–16.

Zeiss, C.J., Ward, J.M. and Allore, H.G. (2012) Designing phenotyping studies for genetically engineered mice. *Veterinary Pathology* **49**, 24–31.

Chapter 3
Gastrointestinal system

Cheryl L. Scudamore

Mary Lyon Centre, MRC Harwell, UK

3.1 Background and development

The gastrointestinal system extends from the oral cavity to the anus and includes the intestines and associated glandular organs. The intestines, liver and spleen arise from the primitive endoderm. The oesophagus, cranial duodenum, stomach and spleen develop from foregut endoderm; liver, gall bladder and pancreas from foregut and midgut junctional endoderm; caudal duodenum, jejunum, ileum, caecum and proximal colon from the midgut endoderm and caudal colon and rectum from the hindgut endoderm (Kaufman and Bard 1999).

Teeth develop from oral ectoderm, which leads to the formation of the epithelial components, such as ameloblasts and the neural crest derived mesenchyme, which results in the formation of odontoblasts and cementoblasts.

The pancreas develops in the embryo from endoderm by the formation of two buds from the ventral and dorsal duodenum. The ventral bud forms the cranial part of the head of the pancreas and the remainder of the pancreas originates from the dorsal bud (Slack 1995). The buds develop into highly branched structures to form the mature pancreas. The exocrine and endocrine cells both appear to arise from endoderm and differentiate during bud development. The endocrine cells are initially individual cells associated with ducts and only form aggregates or islets from day 18.5 of gestation.

3.2 Oral cavity

3.2.1 *Sampling teeth*

Teeth are embedded in the bones of the skull and, as both tissues are calcified, they require decalcification before paraffin-embedded sections can be made or need to be embedded in resin if they are not to be decalcified (Hand 2012; Sterchi 2012). Prior to processing, with or without decalcification, it is important that hard tissues are fixed adequately in formalin for at least 24 to 48 hours. It may be convenient to separate the mandible from the maxilla of the jaw before fixation, but further trimming – producing hemisections of the mandible and maxilla to allow for longitudinal sections – may be best left until the tissue has been softened by decalcification.

For routine examinations teeth are usually decalcified using formic acid or ethylenediaminetetraacetic acid (EDTA)-based techniques. Gooding and Stewart's (10% formic acid and 5% formaldehyde) or Kristesons' declacification fluids have been used for mouse teeth (Mohamed and Atkinson 1983). The end point of decalcification can be tested by a chemical test of the decalcification fluid or by trial incision of the edge of the skull. Decalcification using acid-based techniques may be complete in 2–6 days whereas immersion in EDTA may take a minimum of 2–3 weeks (Seidel *et al.* 2010) before tissues can be trimmed to allow sections can be made. EDTA decalcification is generally preferred if further techniques such as *in situ* hybridization or immunohistochemistry are required.

3.2.2 Anatomy and histology of teeth

Rodent incisors first erupt at days 10–12 and thereafter continue to erupt throughout life growing at a rate of 1 mm/week (Moinichen *et al.* 1996). Rodent incisors therefore have active ameloblast and odondoblast cell layers, which can be seen histologically. Ameloblasts are columnar epithelial cells that line the tooth root in a single layer with basally located nuclei (Figure 3.1). The function of the ameloblasts and the function and nature of their surrounding cell populations changes from the base or secretory region to the maturation region (Josephsen *et al.* 2010). In the secretory region, the ameloblasts are in contact with a layer of flat-to-cuboidal cytokeratin-positive cells, one to four cells thick, the stratum intermedium, which is surrounded by the stellate reticulum and the outer enamel epithelium. The cells of the stellate reticulum are irregular and star shaped with large intercellular spaces. In the maturation zone the

Figure 3.1 Continuously erupting incisor tooth in the mouse has regular layer of ameloblasts (thin arrow). Enamel is often lost during histological processing, leaving an artefactual space between ameloblasts and dentine (star).

ameloblasts are embedded in the papillary layer of irregularly shaped cells, which also contain a rich capillary network. Ameloblasts produce enamel on the labial surface of the incisors throughout life and on all surfaces of the molars during their growth phase, but are not present once the molar teeth have erupted. Enamel is initially produced as a proteinaceous matrix, which is then mineralized and so, in routine decalcified sections, the mineral is removed and mature enamel is not visible leaving an artefactual 'space' in the section – however, basophilic remnants of enamel may be seen in some sections (Figure 3.2). The outer layer of enamel of the incisors also incorporates iron salts but these only reach a high enough concentration to give the characteristic orange appearance to the maxillary incisors in mice (Moinichin *et al.* 1996). Odontoblasts are tall columnar cells with oval nuclei that line the pulp cavity. The nuclei tend to be at varying heights giving the layer a disorganized appearance. Odontoblasts secrete predentine and have a fine apical cytoplasmic process (Tomes fibre), which extends in a tubular space within the predentine layer. The predentine eventually mineralizes. The predentine stains more palely basophilic than the mature dentin (Figure 3.2). The pulp cavity contains a loose connective matrix with embedded blood and lymph vessels (Figure 3.2).

Figure 3.2 Incisor tooth showing enamel remnants retained after processing (thick black arrow), ameloblasts (thin arrow), odontoblasts (white arrow), predentine (blue arrow), dentine (red arrow) and pulp cavity (star).

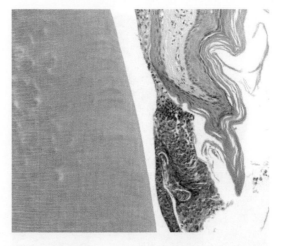

Figure 3.3 Gingival inflammation adjacent to incisor.

The continuous eruption of the incisors means that the process can be disrupted by trauma, nutritional deficiencies and genetic modifications. Disruptions can result in macroscopic changes, including loss of colour, changes in texture and fragility and breakage of the teeth. There is a spontaneous incidence of dental dysplasia of up to 10% in CD1 mice (Losco 1995) and periodontal inflammation is relatively common (Figure 3.3). Minor disruptions in odontoblast and ameloblast function may be detected by irregularities in enamel or dentine deposition (Figure 3.4). Rodent molars first erupt at about day 15 post natal and continue to erupt through the first four months of life. Odontoblasts remain viable within molar teeth, but are metabolically relatively inactive in adult animals (Figure 3.5).

3.2.3 Sampling tongue

The small size of the mouse tongue means that complete longitudinal sections will fit into standard-size tissue cassettes. To ensure that the lingual salivary glands are present in the section, the tissue should be cut lateral to the midline (Ruehl-Fehlert *et al.* 2003).

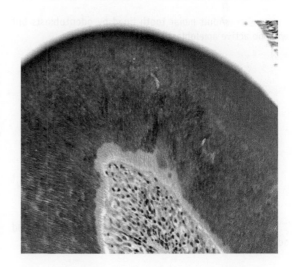

Figure 3.4 Irregularities of predentine and dentine may be early indicators of abnormal enamel formation.

3.2.4 Anatomy and histology of tongue

The tongue is covered by papillae, covered in a stratified squamous epithelium that is keratinized on the dorsal surface (Figure 3.6). Taste buds consist of outer supporting sustentacular cells and the inner elongate chemoreceptor (taste) cells (of which there are three types) (Figure 3.7) connected to nerve

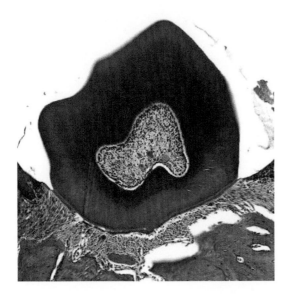

Figure 3.5   Adult molar tooth lined by odontoblasts but with no active ameloblasts.

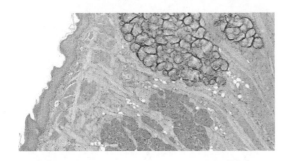

Figure 3.6   Caudal dorsal tongue showing keratinized squamous epithelium and mucous and serous glands.

fibres. The taste buds are predominantly present in the caudal circumvallate papillae of the tongue although they may also be present at other sites in the oral cavity (palate) and pharynx (Treuting and Morton 2012). Serous (von Ebner's glands) and mucous salivary glands are present in the caudal part of the tongue, with ducts emptying on to the dorsal surface (Figure 3.6). The core of the tongue is composed of skeletal muscle arranged in longitudinal and transverse bundles. The muscle is primarily of a fast-twitch type, but the predominant myosin heavy chain varies between muscle bundles (Abe *et al.* 2002). Heavily granulated mast cells are

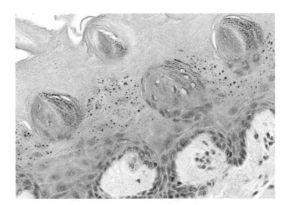

Figure 3.7   High-power photomicrograph of chemoreceptors in dorsal tongue epithelium

commonly seen in the connective tissue between the muscle bundles (Figure 3.8).

3.3   Salivary glands

3.3.1   *Sampling salivary glands*

The mouse has three paired major (encapsulated) salivary glands, the largest of which is the submandibular gland, which is a prominent feature subcutaneously in the ventral neck. All the glands are composed of lobules and have a tubuloalveolar structure. The sublingual and parotid glands lie lateral to the submandibular glands with the sublingual gland in the more cranial portion. The submandibular lymph node, although usually cranial to the salivary glands, can be rather variable in position particularly if altered in size. It may be difficult macroscopically to distinguish and therefore to individually dissect the tissue of the major salivary glands and submandibular lymph node, so it is often convenient to embed the whole complex of tissue and create a single section (Ruhel-Fehlert *et al.* 2004). The mouse also has a number of minor salivary glands which are not distinct encapsulated structures like the major salivary glands but are embedded in their associated tissues. They are therefore usually sampled incidentally with the tissues they are associated with – for example, tongue, pharynx, palate.

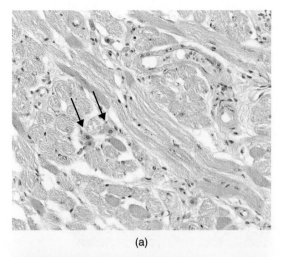

(a)

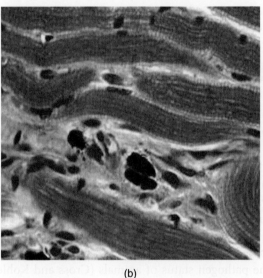

(b)

Figure 3.8 (a) Heavily granulated mast cells (arrows) are commonly seen in normal rodent tongues and can be demonstrated with (b) Toluidine Blue staining.

3.3.2 Anatomy and histology of the salivary glands

All salivary glands are composed of 'secretory end pieces' which are acinar or tubular arrangements of glandular epithelium which empty via intercalated and striated (intralobular) ducts into excretory (interlobular) ducts ending ultimately in the oral cavity (Miletich 2010). In the submandibular gland of the mouse the secretory acinar cells are pyramidal with a mixed seromucous, basophilic cytoplasm and basally located nuclei. There is testosterone-dependent sexual dimorphism in the submandibular glands of mice. In males the acinar cells may contain more zymogen granules and more strikingly the cytoplasm of the convoluted ducts (which connect the intercalated and striated ducts) contains much more prominent eosinophilic granules (Figure 3.9 and Figure 3.10).

The sublingual gland is composed of large columnar mucous-secreting cells, which have pale basophilic cytoplasm (in H&E sections) containing

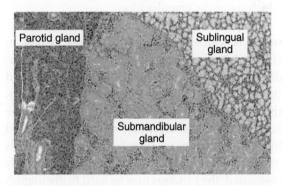

Figure 3.9 Male salivary glands with prominent eosinophilic granulation of convoluted ducts of the submandibular gland.

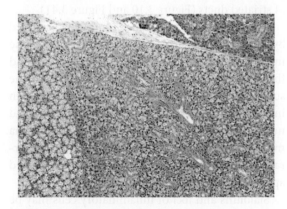

Figure 3.10 Female salivary glands without granuation of convoluted ducts.

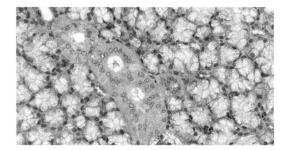

Figure 3.11   Striated ducts of sublingual gland.

large granules that compress the nuclei basally (Figure 3.11). Mucous cells can be demonstrated with alternative stains, for example Alcian blue, which stains the granules blue.

The parotid gland is composed of serous secretory cells, which have a darker basophilic cytoplasm (Figure 3.9). The parotid gland is histologically similar in appearance to both the lacrimal gland and the pancreas. Due to their close anatomic proximity, the lacrimal and parotid glands may appear in the same section but theoretically can be distinguished by the larger amount of cytoplasm in the acinar cells of the lacrimal gland. Although not anatomically co-located, the pancreas may be hard to distinguish from salivary tissue when in fixative so these tissues may be mistaken during the selection and embedding process. Pancreatic tissue can be identified by the presence of islets of Langerhans and the absence of striated ducts (Figure 3.10 and Figure 3.41).

In all of the salivary glands, the ducts are lined by simple cuboidal to columnar epithelial cells. The intercalated duct cells have a low cuboidal lining and these may be hard to appreciate, particularly in the submandibular gland. Striated ducts are often seen in circular cross section and have apically arranged nuclei due to basal invaginations of the plasma membrane, which enclose vertically arranged arrays of mitochondria and give rise to the striated appearance (Figure 3.11).

Sialoliths and foci of inflammatory cells may be found incidentally in the salivary glands of ageing mice (Figure 3.12).

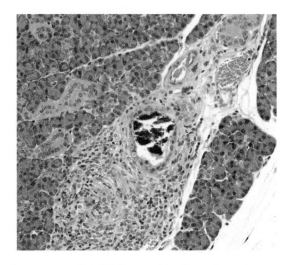

Figure 3.12   Sialolith in parotid salivary duct with associated inflammation.

3.4   Stomach and intestines

3.4.1   Sampling stomach and intestines

The presence of digestive enzymes, bile, acid, digesta and commensal bacterial flora within the intestinal tract means that it is prone to rapid autolysis. The timing of autolysis is variable and can depend on a number of intrinsic and extrinsic factors such as the environmental temperature, pyrexia and even the pathogen status of animals (Cross and Kohler 1969). Rapid fixation of the intestines is therefore very important. Immersion fixation soon after death is often adequate but injection of fixative into the lumen may be helpful in preventing changes and removing intestinal contents.

The mouse stomach wall is very thin, which can make consistent sampling difficult. Whichever method is used it is important to be able to examine the glandular and nonglandular stomach including the transitional region (limiting ridge) and the pylorus. Opening the stomach along the greater curvature and pinning it flat at necropsy (Ruehl-Fehlert et al. 2003) provides good fixation and allows easy orientation of the stomach to provide longitudinal

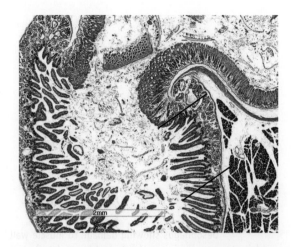

Figure 3.13 Short length of the duodenum demonstrated in subgross image. Duodenum is characterized by Brunner's glands, which are present between the two black lines.

Figure 3.14 Tissue spacers can be used to keep intestines separate in cassettes during processing.

samples through the major regions (mainly from the lesser curvature as this is in the mid line of the pinned tissue). However the small size of the mouse stomach allows for fixation of the organ as a whole, bisection along its length after fixation and embedding one complete half of the stomach, allowing examination of tissue from the greater and lesser curvatures.

As it is not easy to distinguish the different anatomical regions of the small intestine microscopically, it is important to make sure that the small intestine is sampled in a consistent way and that samples can be tracked through to the slide. The duodenum (Figure 3.13) is very short (approximately 2 mm in the mouse) and so should be sampled close to the pylorus; the jejunum can be confidently sampled by taking a section mid-way along the small intestine and the ileum can be sampled within 2 cm of the caecum. Peyer's patches can usually be seen through the serosa (Chapter 1, Figure 1.29) allowing for sampling if necessary. As the anatomical appearance of the intestines can be altered by pathology, it is helpful to ensure that samples are placed into tissue cassettes for processing in such a way that the samples can later be accurately identified, for example by using tissue spacers, placing the tissues in a repeatable pattern (Figure 3.14) or by separating distinct areas of intestine into different blocks.

Cross-sectional samples of oesophagus and intestine often give the best representation of villus architecture without stretching and twisting the tissues, which can lead to distortion – for example artefactual flattening of the mucosa in longitudinal sections (Figure 3.15). However, the reliance on cross-sections may mean that only small regions of the intestine are examined and, as intestinal lesions may be focal or segmental, examination of longitudinal sections may be required. The Swiss-roll technique (Moolenbeek and Ruitenberg 1981) provides another potential solution to this problem by allowing the entire intestine to be fully evaluated, including the gut-associated lymphoid tissue (Elmore 2006). The whole intestine can be rolled in one length or the small and large intestines can be dissected and rolled separately. This method

Figure 3.15 Artefactual flattening of mucosa.

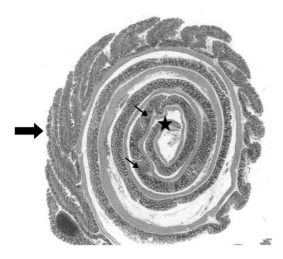

Figure 3.16   Example of Swiss roll of large intestine from caecum (thick arrow) on outside to rectum (star). Longitudinal folds in distal colon can be appreciated (thin arrow).

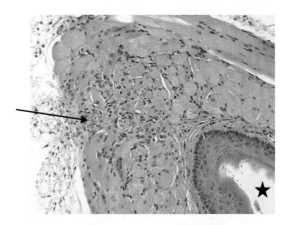

Figure 3.17   Oesophagus showing damage to muscle wall secondary to gavage injury (arrow). Lumen (star) surrounded by keratinized stratified squamous epithelium.

is technically challenging in mice and therefore usually reserved for detailed investigations of the intestine (Figure 3.16).

3.4.2   Anatomy and histology of oesophagus

The oesophagus is lined by a heavily keratinized, stratified squamous epithelium (Figure 3.17). The thickness of the keratin layer can increase rapidly if the animal has not eaten. In normal animals there is usually no adherent ingesta or bacteria. The keratin layer effectively protects the oesophagus from gavage damage, but dosing accidents may occur where the gavage tube penetrates the oesophageal wall. It is uncommon to find the actual site of penetration (Figure 3.17) but inflammation may be seen adjacent to the oesophagus or organs in the mediastinum (thymus and heart base) and occasionally under the skin of the neck or thorax. The lamina propria and muscularis mucosae are indistinct in the proximal oesophagus, becoming more obvious towards the stomach. Throughout its length, the oesophagus is surrounded by two layers of skeletal muscle (inner circular and outer longitudinal) and an outer connective tissue serosa,

although some smooth muscle may be present in the developing embryo (Cao *et al.* 2012).

3.4.3   Anatomy and histology of stomach

The stomach of a rodent is divided into two main anatomical regions distinguished by their epithelial lining – glandular and nonglandular (or squamous). The two areas can be distinguished macroscopically with the luminal surface of the nonglandular region (which is the larger part) appearing pale cream and the glandular area pale pink in colour. The two regions are sharply demarcated by the limiting ridge, which can be seen as a distinct projecting fold. The squamous portion of the stomach extends from the entrance of the oesophagus to the limiting ridge and is lined by a squamous keratinized epithelium, which is normally one to three cells thick. The keratin layer can vary in thickness, for example becoming thicker if the animal has not eaten recently.

Hyperplasia and hyperkeratosis of the nonglandular stomach can be difficult to confirm as the epithelium often becomes folded due to contraction after death. Evidence of mitotic figures above the basal layer and increased cytoplasmic basophilia may help to confirm the diagnosis. At the limiting

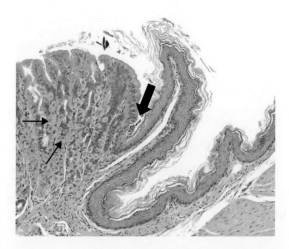

Figure 3.18   Limiting ridge of the stomach with squamous epithelium on the right and glandular epithelium of the cardia on the left. Focal inflammation is common around the limiting ridge (thick arrow) and in the submucosa. Parietal cells are seen as large eosinophilic cells in the glandular stomach mucosa (thin arrow).

ridge the mucosa is folded and there is usually a thick layer of keratin present (Figure 3.18). Intestinal metaplasia of the glandular epithelium (Figure 3.19), focal ulceration and submucosal inflammatory infiltrates adjacent to the limiting ridge are common (Figure 3.18).

The glandular stomach is divided into three regions based on differences in the distribution of cell types in the epithelium. There is a gradual histological transition between the regions. The cardia is a narrow band adjacent to the limiting ridge, the fundus makes up the larger portion of the glandular stomach and the pylorus empties into the duodenum. The glandular epithelium consists of a single layer of cuboidal to columnar cells lining simple, branched gastric pits. In microscopic sections the gastric pits appear straight in mice. In all regions the surface and superficial portions of the pits are lined by surface epithelial cells which are tall columnar with basally placed nuclei and dense, darkly eosinophilic cytoplasm. Mucus-secreting cells are the most abundant cells in the gastric mucosa and are found immediately below the superficial cells in the cardiac and fundic regions, but are most

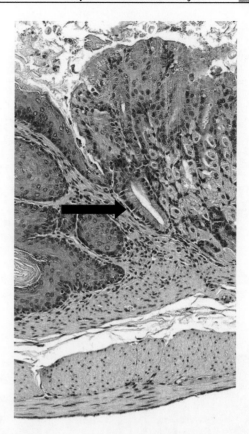

Figure 3.19   Foci of intestinal metaplasia (thick arrow) characterized by columnar epithelial cells with brush border can be seen at the limiting ridge.

prominent in the pyloric region where they line most of the gastric pit (Figure 3.20). Mucus cells are low columnar epithelium with basal nuclei and may have pale, granular mucus containing apical cytoplasm. Parietal cells (acid secreting) are the most visually prominent cell type and are most abundant in the fundic region. They are large cuboidal cells with pale eosinophilic cytoplasm and central round nuclei found scattered between cells along the gastric pits particularly in the fundus (Figure 3.18). Chief (or zymogenic cells) are found at the base of the gastric pits in the fundic region. They are low columnar cells, with dark staining granular eosinophilic cytoplasm and basally placed nuclei (Figure 3.20).

The gastric epithelium covers a connective tissue lamina propria and thin muscularis mucosae, which

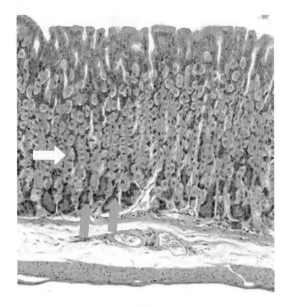

(a)

Figure 3.21 Ulceration of the squamous portion of the gastric mucosa.

is surrounded by connective tissue submucosa and three muscular layers – the innermost oblique, middle circular and outer longitudinal layers. The middle layer tends to be most prominent in mice.

Gastric ulceration can occur as a nonspecific stress response in glandular and nonglandular regions of the stomach (Figure 3.21).

3.4.4 Anatomy and histology of the intestines

The small and large intestines have a similar basic structure to that seen in other mammals with an outermost serosal layer, surrounding outer longitudinal and inner circular smooth muscle layers, submucosa and mucosa. Nerve plexi are present between

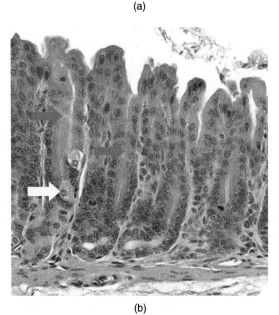

(b)

Figure 3.20 (a) Fundus and (b) pylorus. Paneth cells (thick white arrows) reduce in number towards the pylorus. Chief cells are found towards the base of gastric crypts (green arrows) and mucus cells at the neck of the crypts (blue arrows).

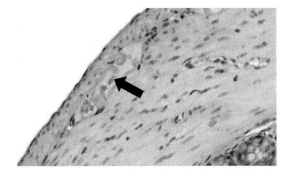

Figure 3.22 Myenteric nerve plexi (arrow) in wall of large intestine between the outer longitudinal and inner circular smooth muscle.

the two muscle layers (Figure 3.22). The lamina propria normally contains variable populations of lymphocytes, plasma cells and macrophages.

In the small intestine the mucosa is thrown up into finger like projecting villi separated by crypts lined by a single layer of columnar epithelial cells, which are predominantly enterocytes with variable populations of mucous, Paneth and neuroendocrine cells along their length. There are three main regions in the small intestine the duodenum, jejunum and ileum. Although the mucosa of these regions have distinct morphological features, in practice they can be hard to distinguish on microscopic sections. The submucosal Brunner's glands are a useful way of confirming that a section is taken from the duodenum (Figure 3.13), although in this region the villi are tallest and 'leaf' shaped in longitudinal section. The jejunal and ileal villi are cylindrical and so have straight sides in section. The jejunal villi are longer and have a greater proportion of Paneth cells in the epithelium compared to the ileum, but this may not be evident on routine sections.

The enterocytes that make up the majority of the epithelial lining of the entire small intestine are tall columnar cells with a brush border and basally located nuclei. The Paneth cells are found in the base of the crypts and contain prominent, large,

eosinophilic granules (Figure 3.23a), which can be stained with a variety of special stains including Sirius red (Figure 3.23b). Paneth cells are more numerous in the jejunum, but can be found along the entire mouse small intestine. Goblet cells are also found along the small intestine but tend to be more numerous in the ileum. Goblet cells have basally located nuclei and large, apical, pale basophilic mucin vacuoles which may be granulated (Figure 3.24). Approximately 1% of the epithelial lining cells are enteroendocrine cells. There are many subtypes distinguished based on their neuroendocrine products and regional distribution (Habib *et al.* 2012).

The large intestine of the mouse is composed of the colon, caecum and rectum. The ileum enters the mouse caecum at the ileocaecal valve and the colon exits from the caecum close to the ileocaecal junction. The colon has distinct regional differences in the appearance of the mucosa with prominent transverse folds proximally, a flat midsection and longitudinal folds distally towards the rectum (Figure 3.16). The caecum is a relatively large, blind-ended sac and also has prominent transverse mucosal folding (Figure 3.16). The caecum and colonic mucosa has deep crypts but no villi (Figure 3.25). The mucosa is lined by epithelium consisting of enterocytes, mucus cells and

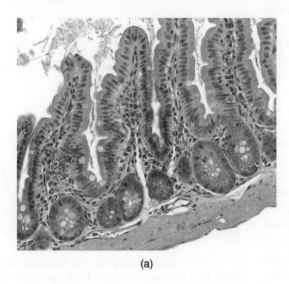

(a)   (b)

Figure 3.23 Jejunum showing Paneth cells staining brightly eosinophilic in crypts with (a) H&E and red with (b) Sirius red.

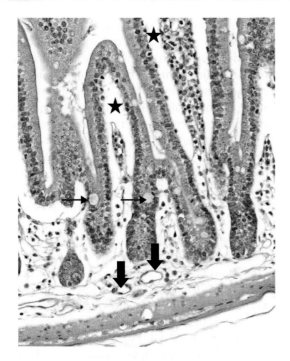

Figure 3.24 Section of ileum with goblet cells (thin arrows). Tissue has been overperfused leading to separation of the epithelium from the core of the villi (stars) and empty distended blood vessels (thick arrows).

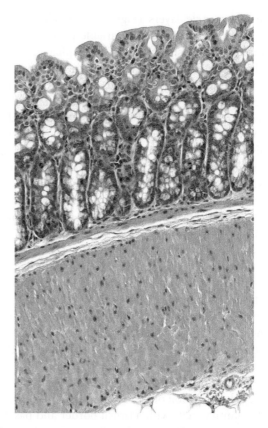

Figure 3.25 Proximal colon with deep crypts and no villi.

enteroendocrine cells (Paneth cells are not present). The mucus cells are more abundant in the large intestine compared to the small intestine and within the large intestine they are more numerous in the proximal colon and caecum than in the distal colon. The rectum is very short in mice and essentially similar in appearance to the distal colon (Figure 3.16, Figure 3.26). There is an abrupt transition to the keratinized squamous epithelium of the anus. Rectal prolapse is relatively common in mice, resulting in mucosal hyperplasia and ulceration.

Gut-associated lymphoid tissue is abundant in mice and is usually located in the anti-mesenteric submucosa. Large organized areas of lymphocytes known as Peyer's patches (Figure 3.27) may be found in the jejunum, ileum and caecum and may be visible from the serosal surface. These tend to be more prominent in young mice. In addition, intraepithelial lymphocytes are common especially in the small

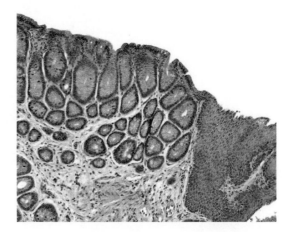

Figure 3.26 Abrupt change at ano-rectal junction from colonic type mucosa with crypts lined by abundant goblet cells and stratified squamous epithelium of the anus.

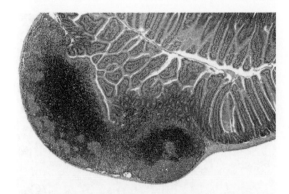

Figure 3.27   Peyers patch in distal small intestine.

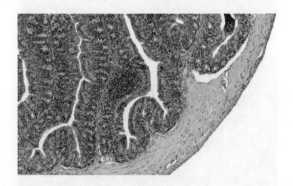

Figure 3.28   Small lymphoid aggregates or 'cryptopatches' can be found anywhere along the length of the intestine.

intestinal mucosa and smaller aggregates of lymphoid cells known as cryptopatches (Figure 3.28) or isolated lymphoid follicles can be found along the length of the intestine (Pabst *et al.* 2005).

Autolysis is a commonly seen artefact in the intestines. It can be seen within a few minutes of death in rodents (Seaman 1987), usually being evident earliest in the duodenum. The earliest change seen is the appearance of a focal subepithelial space between the base of the villous epithelial cells and the lamina propria on the sides or tip of the villi (Cross and Kohler 1969). The space extends until the whole epithelium is separated from the villus while contraction of the lamina propria exacerbates the change (Figure 3.29). The autolytic change should not be confused with overperfusion (Figure 3.24), oedema or dilatation of lymphatics

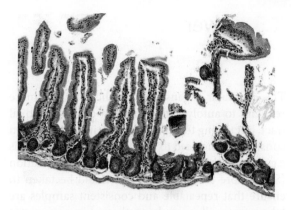

Figure 3.29   Autolysis showing separation from epithelium from villus core and complete loss of epithelium.

(lympangiectasis). The crypt epithelium normally remains attached to the lamina propria during autolysis but eventually the villus epithelium breaks up into individual or groups of cells which slough off into the lumen. Excessive perfusion pressure can also lead to artificial dilatation of blood and lymph vessels and apparent separation of tissues in the lamina propria, which can be confused with oedema.

Pre-weaning mice may have a large number of vacuoles in their intestinal epithelium (Figure 3.30) due to absorption of milk fat which should not be confused with pathological features of rotavirus infection EDIM.

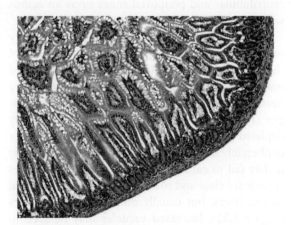

Figure 3.30   Small intestine from pre-weaned mice is usually highly vacuolated.

3.5   Liver

3.5.1   *Sampling liver*

The glandular organs such as the liver are more resistant to autolysis than the rest of the intestinal tract, but timing to fixation may be important when considering the need to perform additional analysis e.g. immunohistochemistry or in situ hybridization (Scudamore *et al.* 2011). Care should be taken to ensure that repeatable and consistent samples are taken from all animals as these organs are not homogeneous in their anatomy or function. The liver lobes have different developmental origins, blood flow and gene expression (Irwin *et al.* 2005). Therefore it is important that a consistent approach is taken, for example routinely taking a sample of left lateral lobe and left and right medial lobes with the gall bladder (Ruehl-Fehlert *et al.* 2003) for histology.

3.5.2   *Anatomy and histology of the liver*

The mouse liver follows the conventional mammalian pattern. It has four lobes (large left lobe, divided right lobe and a caudate lobe) and a gall bladder. Microscopically, the liver can be viewed as a classical lobule centred on the central vein with centrilobular and periportal zones or as an acinus with three functional zones.

The appearance of hepatocytes can vary markedly depending on the nutritional status of the animals. Hepatocytes of normal, unstarved mice have abundant clear cytoplasm which has a 'feathery' appearance due to the presence of glycogen (Figure 3.31). Glycogen stores can be rapidly depleted (Figure 3.31) if animals are starved either deliberately as part of an experimental protocol or if they fail to eat because they are ill. Fat vacuoles (which are clear and round) may be seen in normal mouse livers, but usually only in small numbers (Figure 3.32). Increased vacuoles may be seen in the periportal areas of starved animals as a result of fat mobilization. The use of high fat diets, however, can result in marked changes in liver morphology

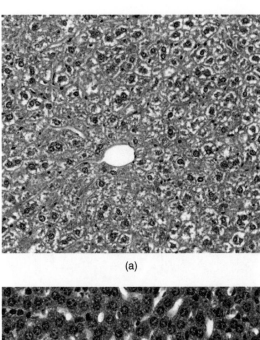

(a)

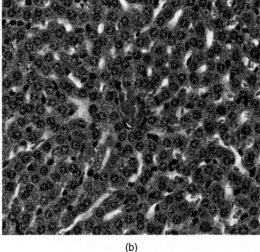

(b)

Figure 3.31   (a) Liver from a normally fed animal, hepatocytes with 'feathery' cytoplasm containing glycogen. (b) Liver from a starved animal, hepatocytes with condensed darkly staining cytoplasm depleted of glycogen.

with the production of both microvesicular and macrovesicular vacuolation (Figure 3.33). These changes can make it hard to interpret more subtle treatment-related changes in the liver and so the necessity of incorporation of high fat diets into protocols should be carefully evaluated if morphological endpoints are anticipated in the liver.

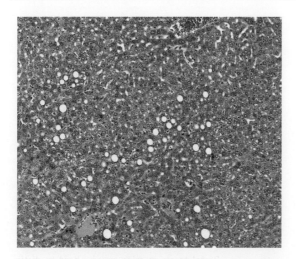

Figure 3.32 Small numbers of clear fat containing vacuoles may be seen in normal livers and may increase in animals that have been fasted.

There are a number of features that are commonly found in the mouse liver and are considered part of the normal back ground histology and so may not be related to treatment or genetic modification. However in certain circumstances these background changes may increase or decrease and it is important to recognize that this may be related to manipulation of the mouse or its environment.

These background lesions include small foci of inflammatory cells that are commonly found in mouse livers and usually have no significance and may reflect showering of the liver with antigens absorbed from the intestines. These cells are usually mononuclear (lymphocytes and macrophages) but may contain occasional neutrophils. Occasionally, they may also be associated with single cell necrosis. Larger foci of coagulative necrosis may also be seen but multiple foci and/or necrosis covering a large area should be considered potentially abnormal. The liver is a common site for extramedullary haematopoiesis (Figures 3.34 and 3.35) in the mouse along with the splenic red pulp. Haematopoietic cells may be predominantly small dark-staining nucleated red-cell precursors or large granulocyte precursors with band or ring nuclei. In addition occasional megakaryocytes may be present scattered

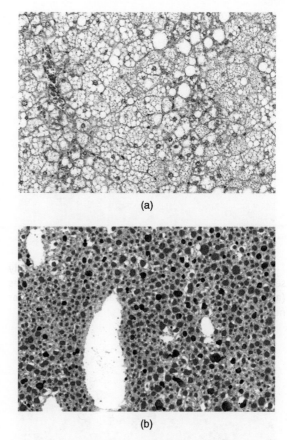

(a)

(b)

Figure 3.33 Animals on a high-fat diet may have variable levels of micro and macrovesicular vacuolation ((a) H&E stained) which can obscure other potential phenotypic changes in morphology. Fat vacuoles can be demonstrated in frozen sections using Oil-Red-O or similar special stains (b).

in the hepatic parenchyma which should not be confused with unusual mitotic figures or hyperplastic hepatocytes (Figure 3.34). Haematopoiesis is present normally in neonatal mice (Figure 3.35) and may be increased in response to any condition which generates anaemia.

It is common for mouse hepatocytes to be binucleate and also for hepatocyte polyploidy to increase with age (Fagiolli *et al.* 2011) This can result in enlarged hepatocytes with enlarged nuclei (karyomegaly) or multiple nuclei (Figure 3.36). Portal areas/tracts may not be very obvious in mouse

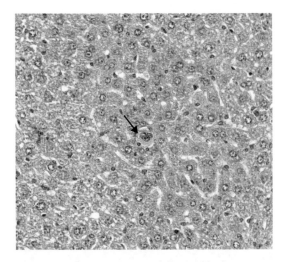

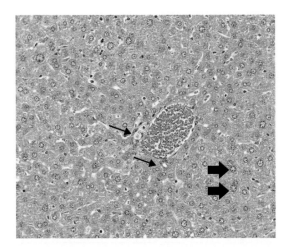

Figure 3.34   Extramedullary haematopoiesis is commonly seen in mouse livers and can include megakaryocytes (arrow), which should not be mistaken for polyploid or neoplastic cells.

Figure 3.36   Portal tracts in mice may be hard to distinguish and only one or two cross sections of bile duct (thin arrows) may be present. Variable sized nuclei as seen in this section and binucleate nuclei (thick arrow) are common and increase with age in mice.

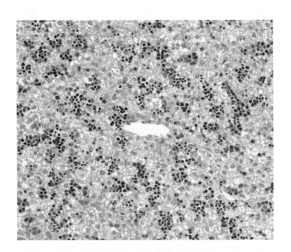

Figure 3.35   Neonatal liver showing extensive deposits of extramedullary haematopoiesis predominantly dark staining erythrocyte precursors.

livers due to small amounts of connective tissue (Figure 3.36). Usually only one or two cross sections of bile duct will be seen in the portal area and it may be hard to appreciate the other components, which should be present, namely branches of the hepatic portal vein, hepatic artery and lymphatics. Bile

canaliculi, which run between hepatocyte plates, and feed into bile ducts via the ducts of Hering are also difficult to appreciate in normal sections. Kupffer cells or resident macrophages are a common feature of the endothelial lined sinusoids, which lie between cords of hepatocytes and can be highlighted using immunohistochemical markers for example F4/80. Ito or stellate cells are found in the space of Disse between the sinusoidal endothelium and hepatocytes. Ito cells have a number of functions including vitamin A and fat storage, antigen presentation and collagen production during fibrosis (Winau *et al.* 2007) and can be identified using GFAP immunohistochemistry. Other resident cell populations in the liver include T and NK cells.

The gall bladder is lined by cuboidal epithelial cells a proportion of which are brush cells which have microvillus surface projections (Luciano and Reale 2007). The epithelium is supported by a lamina propria of loose connective tissue which often contains scattered inflammatory cells, and an external smooth muscle surrounded by connective tissue adventitia (Figure 3.37).

Artefactual cracking of the central portion of a liver lobe cross section usually indicates that the

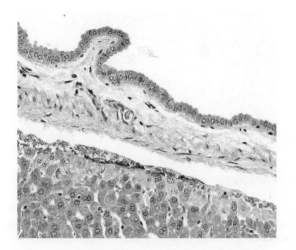

Figure 3.37   Gall bladder lined by cuboidal epithelium.

liver has not been fixed for long enough in formalin. Formalin fixation takes a minimum of 24 hours. If tissues are processed too soon the central area of the tissue will not be properly fixed with formalin and as a result, secondary fixation with alcohol occurs during in the tissue processor. The central alcohol fixed tissue contracts more than surrounding formalin fixed tissue resulting in increased tension on the tissue and cracking (Figure 3.38).

3.6   Pancreas

3.6.1   *Sampling pancreas*

Within the pancreas the number and size of islets of Langerhans varies between regions (Fig 3.39), with physiological state, disease (diabetes) and with age (Adeghate *et al.* 2006). Larger islets tend to be seen close to nutrient arteries and ducts, so consistent sampling is important to avoid overinterpretation of variation in islet size (Kilimnik *et al.* 2009; Hornblad *et al.* 2011). For routine analysis sampling of the left or splenic lobe of the pancreas which is present adjacent to the spleen may be convenient with the remainder of the pancreas sampled with the duodenum and small intestine if required (Ruehl-Fehlert *et al.* 2003). The left lobe should be embedded so that a longitudinal section is made. For more detailed analysis and morphometry the entire pancreas should be carefully dissected free and laid flat on filter paper prior to fixation (Kilimnik *et al.* 2009).

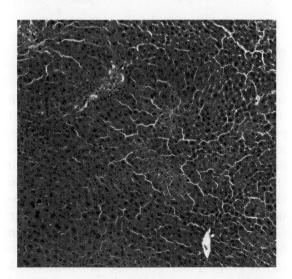

Figure 3.38   Insufficient fixation time in formalin can result in cracking of liver tissues as a result of secondary fixation with alcohols during processing.

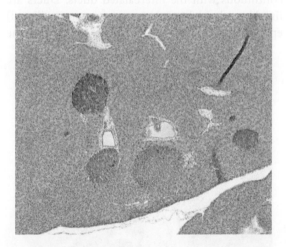

Figure 3.39   Pancreas sample stained using immunohistochemistry to detect insulin. Brown staining highlights insulin staining cells of the islets and demonstrates the normal variation in islet cell size. Separation between pancreatic lobules is commonly seen in sections from normal mice.

3.6.2 Anatomy and histology of the pancreas

The pancreas consists of three lobes – splenic, duodenal and gastric – which contain two major components, the exocrine secreting acinar tissue and the endocrine cells located in the islets of Langerhans. The pancreas is surrounded by loose connective tissue of the mesentery and divided into lobules by connective tissue septae. Artificial separation of lobules is a common artefact and needs to be distinguished from generalized oedema which is often manifest by wide separation of lobules (Figure 3.39, Figure 3.42). The lobules consist of the exocrine cells, which are clustered in acini and are drained centrally via intercalated ducts. The duct system drains via intralobular and interlobular ducts to the main pancreatic duct, which empties into the duodenum via the common bile duct (Figure 3.40). The acinar cells are roughly triangular in shape with basophilic cytoplasm and usually two basally located nuclei. The apical portion of the cells contains multiple prominent, discrete, highly eosinophilic zymogen granules (Figure 3.41). Luminally, the acinar cells are lined by flattened pale staining centroacinar cells which are continuous with the intercalated ducts. Ducts are lined by cuboidal to columnar (sometimes ciliated) epithelial cells with occasional goblet cells in the

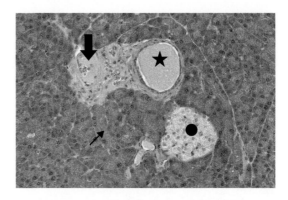

Figure 3.41 Pancreas showing zymogen granules (thin arrow), pancreatic duct (star) and associated blood vessel containing RBCs (thick arrow) with islet (circle). H&E.

Figure 3.42 Islet cell hyperplasia is common in older mice.

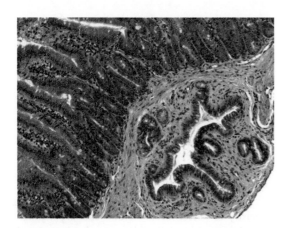

Figure 3.40 The common bile duct empties into the duodenum and shouldn't be mistaken for pathology.

larger diameter ducts. The islets are dispersed within the exocrine pancreas and composed of cords of pale staining, polygonal cells with central nuclei surrounding a capillary network. The islets are predominantly composed of insulin producing cells (83%) (Figure 3.39) with lesser proportions of glucagon (11%), somatostatin (10%) and pancreatic polypeptide (PP) (5%) secreting cells (Aldehgate *et al.* 2006). Glucagon, somatostatin and PP secreting cells can be demonstrated using IHC and tend to be localized to the periphery of the islets (Aldhegate *et al.* 2006). Islets can be variable in size (Figure 3.39) and hyperplasia is common in ageing mice (Figure 3.42).

References

Abe, S., Maejima, M., Watanabe, H. *et al.* (2002) *Muscle-fiber characteristics in adult mouse-tongue muscles Anatomical Science International* 77, 145–148.

Adeghate, E., Howarth, C., Rahsed, H. *et al.* (2006) The effect of a fat-enriched diet on the pattern of distribution of pancreatic islet cells in the C57BL/6J mice. *Annals of the New York Academy of Sciences* **1084**, 361–370.

Cao, X.-M., Yang, Y.-P., Li, H.-R. *et al.* (2012) Morphology of the developing muscularis externa in the mouse esophagus. *Diseases of the Esophagus* **25**, 10–16.

Cross, R.F. and Kohler, E.M. (1969) Autolytic changes in the digestive system of germfree, Escherichia coli monocontaminated, and conventional baby pigs. *Canadian Journal of Complementary Medicine* **33**, 108–112.

Elmore, S.A. (2006) *Enhanced histopathology of mucosa-associated lymphoid tissue Toxicologic Pathology* **34** (5), 687–696.

Faggioli, F., Vezzoni, P. and Montagna, C. (2011) Single-cell analysis of ploidy and centrosomes underscores the peculiarity of normal hepatocytes. *PLoS ONE* **6** (10), e26080. doi:10.1371/journal.pone.0026080

Habib, A.M., Richards, P., Cairns, L.S. *et al.* (2012). Overlap of endocrine hormone expression in the mouse intestine revealed by transcriptional profiling and flow cytometry. *Endocrinology* **153**, 3054–3065.

Hand, N.M. (2012) Plastic embedding for light microscopy, in *Theory and Practice of Histological Techniques*, 7th edn (eds K.S. Survana, C. Layton and J.D. Bancroft), Churchill Livingstone, Edinburgh.

Hornblad, A., Chedadd, A. and Ahlgren, U. (2011) An improved protocol for optical projection tomography reveals lobular heterogeneities in lobular pancreatic islet and β cell mass distribution. *Islets* **3**, 204–208.

Irwin, R.D., Parker, J.S., Lobenhofer, E.K. *et al.* (2005) Transcriptional profiling of the left and median liver lobes of male F344/N rats following exposure to acetaminophen. *Toxicologic Pathology* **33**, 111–117.

Josephsen, K., Takano, Y., Frische, S. *et al.* (2010) Ion transporters in secretory and cyclically modulating ameloblasts: a new hypothesis for cellular control of preeruptive enamel maturation. *American Journal of Physiology – Cell Physiology* **299**, C1299–C1307.

Kaufman, M.H. and Bard, J.B.L. (1999) *The gut and its associated tissues, in* The Anatomical Basis of Mouse Development, Academic Press, San Diego, CA.

Kilimnik, G., Kim, A., Jo, J. *et al.* (2009) Quantification of pancreatic islet distribution *in situ* in mice. *American Journal of Physiology, Endocrinology and Metabolism* **297**, E1331–E1338.

Losco, P.E. (1995) Dental dysplasia in rats and mice. *Toxicologic Pathology* **23**, 677–688.

Luciano, L. and Reale, E.. (1997) Presence of brush cells in the mouse gallbladder. *Microscopy Research and Technique* **15**; **38** (6), 598–608.

Miletich, I. (2010) Introduction to salivary glands: structure, function and embryonic development, in *Salivary Glands: Development, Adaptations and Disease* (eds A.S. Tucker, and I. Miletich), Karger, Basel, pp. 1–20.

Mohamed, S.S. and Atkinson, M.E. (1983) A histological study of the innervation of developing mouse teeth. *Journal of Anatomy* **136**, 735–749.

Moinichen, C.B., Lyngstradaas, S.P. and Risnes, S. (1996) Morphological characteristics of mouse incisor enamel. *Journal of Anatomy* **189**, 325–333.

Moolenbeek, C. and Ruitenberg, E.J. (1981) The 'Swiss roll': a simple technique for histological studies of the rodent intestine. *Laboratory Animals* **15**, 57–59.

Pabst, O., Herbrandt, H., Worbs, T. *et al.* (2005) Cryptopatches and isolated lymphoid follicles:dynamic lymphoid tissues dispensable for the generation of intraepithelial lymphocytes. *European Journal of Immunology* **35**, 98–107.

Ruehl-Fehlert, C., Kittel, B., Morawetz, G. *et al.* (2003) Revised guides for organ sampling and trimming in rats and mice – Part 1. *Experimental and Toxic Pathology* **55**, 91–106.

Scudamore, C.L., Hodgson, H.K., Patterson, L. *et al.* (2011) The effect of post-mortem delay on immunohistochemical labelling – a short review. *Comparative Clinical Pathology* **20**, 95–101.

Seaman, W.J. (1987) *Postmortem Change in the Rat*, Iowa State University Press, Ames, IA.

Seidel, K., Ahn, C.P., Lyons, D. *et al.* (2010) Hedgehog signaling regulates the generation of ameloblast progenitors in the continuously growing mouse incisor. *Development* **137** (22), 3753–3761.

Slack, J.M.W. (1995) Developmental biology of the pancreas. *Development* **121**, 1569–1580.

Sterchi, D.L. (2012) Bone, in *Theory and Practice of Histological Techniques*, 7th edn (eds K.S. Survana, C. Layton and J.D. Bancroft), Churchill Livingstone, Edinburgh.

Treuting, P.L. and Morton, T.H. (2012) Oral cavity and teeth, in *Comparative Anatomy and Histology: A Mouse and Human Atlas* (eds P.M. Treuting and S. Dintzis), Elsevier, Amsterdam.

Winau, F., Hegasy, G., Weiskirchen, R. *et al.* (2007) Ito cells are liver-resident antigen-presenting cells for activating T cell responses. *Immunity* **26**, 117–129.

Chapter 4
Cardiovascular system

Cheryl L. Scudamore
Mary Lyon Centre, MRC Harwell, UK

4.1 Background and development

The cardiovascular system of the mouse follows the standard pattern for most mammalian species and comprises the heart, blood vessels and lymphatics. The heart in all mammalian species has striking similarities and given the extensive use of mice to model cardiac disease phenotypes it is helpful that the mouse heart is very similar morphologically to the human heart, with the exception of the arrangement of some of the venous vessels entering the atria (Wessels and Sedmera 2003). The heart is the first organ to develop in the embryo and is formed from cardiomyocytes and endothelial cells originating from cardiogenic mesoderm. The cardiomyocytes and endothelium initially form tubes which loop and eventually divide into the four chambers (Kaufman and Bard 1999; Savolainen *et al.* 2009). Additional cell types are incorporated into the final heart structure and include differentiation of epicardium, aortic arches and valves from neural crest cells, coronary arteries from mesenchyme and atrial septum from mesocardium.

4.2 Sampling techniques and morphometry

Mouse heart weights vary between strains (Roth *et al.* 2002; Sugiyama *et al.* 2002; Krinke 2004) and although data are available in the literature comparison to concurrent controls (or reference data from the same laboratory) is preferred when investigating experimental variations. Absolute weights are rarely helpful and although comparison of organ to body weight is frequently used, this may not be appropriate where body weight changes are expected. The relationship to brain weight may also not be linear and so other methods of comparison, such as covariance analysis, may be required where subtle changes are suspected (Bailey *et al.* 2004).

The small size of the mouse heart means that it can be conveniently sectioned longitudinally to

A Practical Guide to the Histology of the Mouse, First Edition. Cheryl L. Scudamore.
© 2014 John Wiley & Sons, Ltd. Illustrations © Veterinary Path Illustrations, unless stated otherwise.
Published 2014 by John Wiley & Sons, Ltd. Companion Website: www.wiley.com/go/scudamore/mousehistology

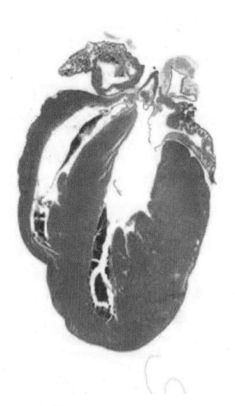

Figure 4.1   Longitudinal section of mouse heart demonstrating all the cardiac chambers, valves and major vessels at the heart base.

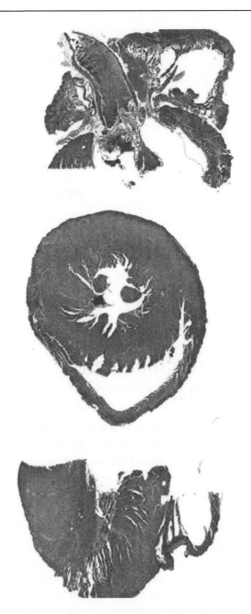

Figure 4.2   Alternative sectioning of heart showing transverse section through ventricles and longitudinal sections of base and apex.

show all chambers and valves (Morawietz *et al.* 2004) (Figure 4.1). While this appears to be a simple technique, in practice the small size may make orientation difficult and it is therefore important to practise and standardize the technique to ensure consistent sampling. It is important to be aware of which anatomical regions are of most interest to ensure these are sampled accurately – for example, the papillary muscle is susceptible to toxicity from a number of drug classes and is easily missed from sections. In some circumstances it may be useful to use the three-section technique (Isaacs 1998), where a cross section is taken from the middle of the ventricles and longitudinal sections of the base and apex of the heart (Figure 4.2). This technique allows more accurate comparison and morphometry of

the relative thickness of the interventricular septum and ventricular walls and can be helpful to optimize presentation of the valves. With modern imaging techniques (microCT – Martinez *et al.* 2009 and echocardiography – Liu and Rigel 2009) detailed morphometric techniques can be made from the heart *in vivo*. However if these techniques are not available it may be necessary to perform morphometry on fixed sections. If this is the case then the heart should ideally be stopped at a consistent stage of the cardiac cycle to avoid the confounding effects of contraction or relaxation on the apparent thickness of the myocardium during systole and diastole. This is normally achieved by the injection of potassium chloride to stop the heart in diastole.

4.3 Artefacts

4.3.1 *Perfusion artefacts*

Perfusion fixation is commonly used in mice to optimize tissue morphology, particularly in the CNS, and to facilitate subsequent techniques such as immunohistochemistry. The most commonly used perfusion techniques for the CNS involve infusion of fixative directly into the left ventricle or into the ascending aorta via the left ventricle. Perfusion via the left ventricle often leads to artefactual distension of the cardiac chambers particularly if the pressure is too high (Figure 4.3). The distension and subsequent thinning of the ventricular walls, which should not be confused with pathological thinning or atrophy of the ventricular walls such as may be seen in dilated 'cardiomyopathy'. Other routes for perfusion can be considered and include perfusion via the descending aorta or vena cava for whole animal perfusion, via the descending aorta proximal to the distal bifurcation for the kidney and the portal vein for the liver. Even optimized perfusion also leads to changes in the appearance of routinely stained tissue due to the absence of blood cells within the blood vessels (Figure 4.4). The tissue can look strikingly different from tissue that has not been perfused and this can lead to variation within experiments where some animals are well perfused and others are not. Where the perfusion pressure is too high this can lead to disruption of tissue architecture and artificial separation of tissue structures, which may be easily confused with oedema (Figure 4.5).

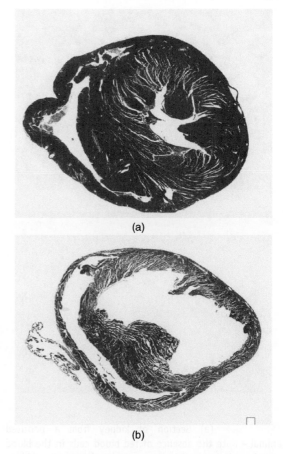

(a)

(b)

Figure 4.3   (a) Transverse section of nonperfused heart demonstrates the thick walls of the left ventricle and the interventricular septum and the thin right ventricular wall. (b) Transverse section demonstrating the artefactual distension of left and right cardiac chambers and thinning of the wall due to intracardiac perfusion at necropsy.

4.3.2 *Contraction artefacts*

A number of artefacts can be introduced to cardiac muscle during handling at necropsy. Cutting the muscle followed by rapid immersion fixation of reactive or beating cardiac muscle can result in artefactual contraction bands (Figure 4.6), which may mimic contraction band necrosis (Olmesdahl *et al.* 1979). Crushing, cutting and handling can also result in the formation of scattered irregular groups of hypereosinophilic fibres (Figure 4.7)

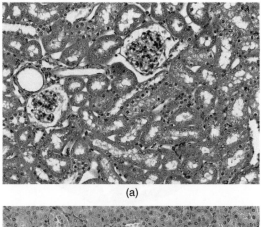

(a)

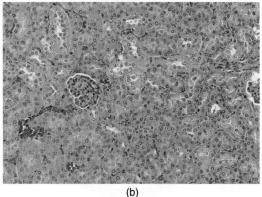

(b)

Figure 4.4 (a) Section of kidney from a perfused animal – note the absence of red blood cells in the blood vessels and glomeruli in comparison to (b) showing kidney from an unperfused mouse.

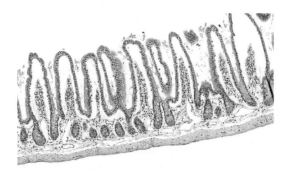

Figure 4.5 The effect of overperfusion is shown in this section of jejunum where there is distension of the blood vessels and separation of the tissues giving a false impression of oedema.

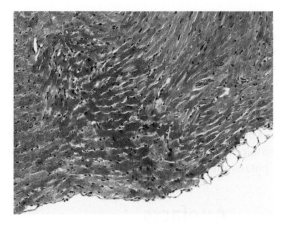

Figure 4.6 Contraction band artefact in cardiac muscle can occur due to handling and fixation at necropsy.

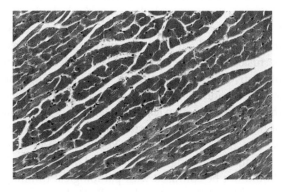

Figure 4.7 Artefactual eosinophilic fibres need to be distinguished from degeneration. True degenerative fibres may show fragmentation and vacuolation as well as eosin-ophilia and are normally accompanied by an inflammatory response.

(Keenan and Vidal 2006). These hypereosinophilic fibres should be differentiated from the earliest signs of cardiomyofibre degeneration. Cardiomy-ofibre degeneration will occur within a few hours of insult therefore if hypereosinophilic fibres are the only change present they are most likely to repre-sent an artefact rather than a genuine pathology. Immersion of the heart into chilled normal saline or formalin before incision may help to reduce artefacts.

4.4   Anatomy and histology of the heart

The heart is surrounded by the pericardial sac, which is lined by mesothelium. In most routine sections only the visceral pericardium, the mesothelial layer lining the outside of the heart, is visible. Normal mesothelial cells are very thin, squamous cells that originate from the mesoderm. They form a thin monolayer over a thin basement membrane, which lines the body cavities and produce a lubricating fluid secretion. If the mesothelial cells are perturbed they tend to become cuboidal. Frond-like projections of fibrous connective tissue with a mesothelial covering can sometimes be found incidentally in the pleura and pericardium.

The epicardium consists of the mesothelial covering and a thin layer of connective tissue immediately below it, which also contains blood vessels and fat, particularly at the heart base. Blood vessels can be seen running between myocardial fibres. The mouse generally has two coronary arteries arising from the left and right aortic sinuses supplying the left and right ventricles respectively. These arteries predominantly occur within the myocardium. The interventricular septum is supplied by septal arteries, which may arise from left or right coronary arteries (Fernández *et al.* 2008). The actual pattern of coronary arteries can be quite variable and so care should be taken when interpreting possible phenotypic changes (Fernández *et al.* 2008).

Beneath the epicardium is the main muscle mass of the heart, the myocardium. The endocardial surface of the ventricles is lined by branching longitudinal folds known as the trabeculae carnae. Unlike in man, there is no clear difference in the width of these trabeculae between the left and right ventricles (Wessels and Sedmera 2003). The myocardium consists predominantly of cardiac myocytes, which are cylindrical, branching fibres that are joined end-to-end to adjacent cells by intercalated discs (Figure 4.8). The intercalated discs are composed of adherens junctions and desmosomes, which help to maintain contact between adjacent fibres during contraction (Li and Radice 2010) and gap junctions

Figure 4.8   PTAH stained cardiac muscle demonstrating the intetcalated discs (pale pink bands between adjoining fibres) and striations.

which allow transmission of impulses between adjacent fibres. Cardiacmyocytes have one or two centrally placed nuclei. Enlarged cardiomyocyte nuclei are common in mouse hearts (Elwell and Mahler 1999), particularly in the papillary muscle (Figure 4.9), and may indicate polyploidy or a response to injury (Keenan and Vidal 2006; Liu *et al.* 2010). The arrangement of the contractile proteins within the myofibre leads to the striated appearance. The presence (or absence) of striations can be

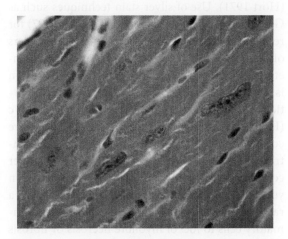

Figure 4.9   Enlarged nuclei can often be found in the papillary muscle and interventricular septum of control mice.

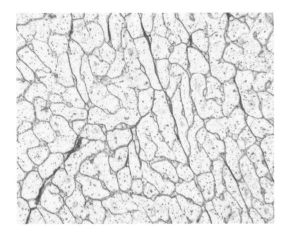

Figure 4.10   Reticulin stains can be useful to high light the borders of myofibres for morphometry.

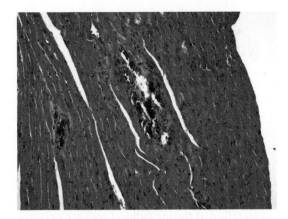

Figure 4.11   Focal mineralization of the ventricles is a common incidental finding in some strains of mice. This spontaneous finding needs to be distinguished from mineralization of necrotic tissue secondary to induced cardiac damage.

highlighted using special stains, for example phospho tungstic acid haematoxylin (PTAH) (Figure 4.8) and antibodies to cardiac specific proteins like troponin can be used to demonstrate early signs of myocyte damage. Cardiac myocyte fibres are arranged in a spiral pattern around the heart, so elliptical cross sections are most commonly seen, making it hard to measure myofibre length. Cross-section width can also be hard to measure consistently therefore measurements of cross-sectional area may be more reproducible for morphometry (Hort 1971). Use of silver stain techniques such as Gordon and Sweet's (Bishop and Louden 1997) for reticulin fibres can be useful to delineate the borders of fibres to facilitate image analysis and morphometry (Figure 4.10). Calcification of the myocardium is a relatively common spontaneous finding particularly in certain mouse strains, such as BALB/c (Figure 4.11). It can occur from a young age, is more common in males and usually affects the right ventricle (Frith and Ward 1988; Taylor 2011). Another common spontaneous finding that can be seen in young mice is 'cardiomyopathy', which can have a range of presentations from small clusters of inflammatory cells to foci of myofibre necrosis or fibrosis (Taylor 2011). Comparison of the incidence and severity of these changes with concurrent controls or recent historical controls from the same strain

maintained in the same facility is important to establish their significance. Atrial thrombosis is also fairly frequently observed in older mice with significant differences between strains (Taylor 2011).

The cardiac conduction system arises from the myocardial cells (Christoffels and Moorman 2009) and, as in other species, comprises the sinoatrial node, atrioventricular node, bundles of His and Purkinje fibres (Rentschler et al. 2001). Identifying these structures in H&E-stained tissues is difficult but the cells contain connexion 45, which can be detected using immunohistochemistry. The cardiac skeleton is made up of connective tissue, which provides support for the valves and myocardium. The fibrous skeleton forms annuli, which keep the atrioventricular (AV) valve lumena open. The fibrous tissue extends into the AV valves, chordae tendinae, papillary muscle and into the muscle at the heart base, which can lead to a disorganized appearance that should not be confused with fibrosis or cardiomyopathy (Figure 4.12). Islands of cartilagenous metaplasia can occur in the cardiac skeleton at the base of the heart or in the aorta in normal mice (Qiao et al. 1995) (Figure 4.13).

The heart valves (aortic, pulmonic and atrioventricular) are made up of three structural layers covered by endothelium (Figure 4.14). The ventricularis

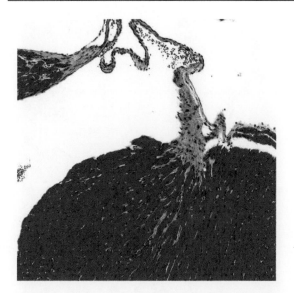

Figure 4.12 The cardiac skeleton at the base of the heart can be demonstrated with stains for fibrous tissue like Masson's trichrome. The normal structure of the cardiac skeleton should not be confused with pathological fibrosis or cardiomyopathy.

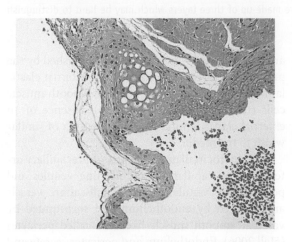

Figure 4.13 Islands of cartilage may be seen in the region of the cardiac skeleton in normal mice and associated with arteritis at the base of the great blood vessels.

is the layer closest to the blood flow and is composed of elastic fibres. The spongiosa, which is composed of collagen and proteoglycan, makes up the core of the valve. The outer layer is the fibrosa, which is composed of dense collagen, which provides structural

stiffness and extends into the chordae tendinae thereby anchoring the valve leaflet. In some mouse strains, for example C57Bl/6, it is common to see black melanin pigment in the valve leaflets particularly in the aortic valve (Hinton *et al.* 2008) (Figure 4.15). It may be necessary to distinguish the melanin pigment from calcification of the valves and this can be done using a calcium-specific stain such as Alizarin red. The pulmonic valve may have granular blue staining in routine H&E stained sections and this is considered a normal feature (Elwell and Mahler 1999).

4.5 Anatomy and histology of the blood vessels

The vascular system of the mouse is essentially similar to other mammalian species. Blood vessels other than capillaries have three layers – the intima, media and adventitia – which are variably developed depending on whether the vessel is an artery or vein and its diameter (calibre). The intima is the innermost layer and is lined by a single layer of endothelium with subendothelial loose connective tissues, some vascular smooth muscle cells and in arteries an internal elastic lamina. The middle layer or media is composed of concentrically arranged smooth muscle, elastic fibres, reticular fibres and proteoglycans. The adventitia usually blends with the outer layer of the media and consists predominantly of collagen and elastic fibres.

Close to the heart, where the blood pressure is highest, the arteries such as the aorta (Figure 4.16) are large-calibre vessels with a high proportion of elastic fibres (elastic arteries) in their walls, which may give them a slightly yellow appearance on gross examination. These arteries also have large amounts of elastic fibres in media which may make it hard to distinguish the internal elastic lamina. The adventitia of these large arteries is relatively undeveloped and in the largest arteries may include small blood vessels, the vaso vasorum, which supply nutrients to the outer wall of the parent vessel. Brown adipose tissue is often found around the aorta and can

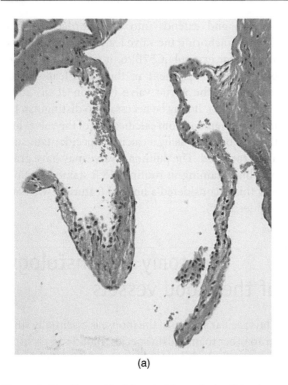
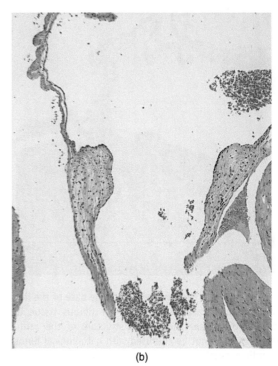

(a)   (b)

Figure 4.14   The aortic (a) and atriovenricular (b) valves are made up of three layers which may be hard to distinguish in H&E sections.

provide a consistent site for assessing this tissue as it is generally not mixed with white adipose tissue and is less likely to be subject to trauma due to handling or micro-chipping than the interscapular brown fat pad (Figure 4.16). Inflammatory changes and mineralization are occasionally seen as apparently spontaneous changes at the base of the aorta (Taylor 2011) as it originates from the heart.

The next calibre of vessels are the muscular arteries, which have a prominent internal elastic lamina, many layers of smooth muscle cells in the media, an obvious external elastic lamina and a well developed adventitia. Where smaller calibre blood vessels branch off from muscular arteries there may be an obvious thickening of the intima (Figure 4.17), which should not be confused with pathological hypertrophy. Small arteries can be distinguished from neighbouring veins by the relative prominence of smooth muscle in their walls compared to that seen in the veins (Figure 4.18). The smallest calibre vessels on the arterial side of the circulation are the arterioles which can be distinguished by the presence of a very thin intima and internal elastic lamina, media which only contains smooth muscle cells (usually less than six layers), absence of an external elastic lamina and an adventitia of similar width to the media.

The microcirculation consists of precapillary arterioles, the capillary bed, collecting venules and postcapillary venules. These small-calibre vessels are all lined by endothelium and surrounded by vascular smooth muscle-like cells called pericytes (Hall 2006). Endothelium and pericytes secrete and share the same basement membrane, unlike vascular smooth muscle cells, which are separated from the endothelium by the intima. A range of markers such as αSMA (smooth muscle actin) and NG2 proteoglycan can be used to label pericytes and vascular smooth muscle cells, but there are no markers that can reliably distinguish the two cell types. The vessels of the microcirculation can largely by identified by light microscopy based on diameter, with

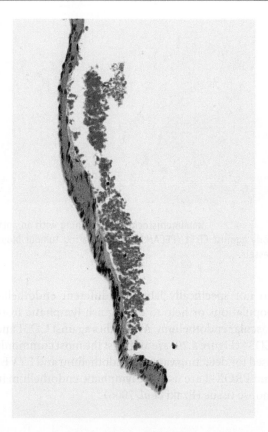

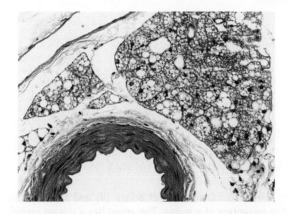

Figure 4.16 Cross section of aorta. Elastic fibres are seen as more darkly eosinophilic wavy lines in the intima and media of the vessel wall. Brown fat is found adjacent to the aorta in normal mice and can be a useful place to assess this tissue if separate samples of brown fat have not been taken from the shoulder region.

Figure 4.15 Melanin pigment deposits are common incidental findings in the valves of pigmented mouse strains and should be distinguished from pathological mineralized deposits.

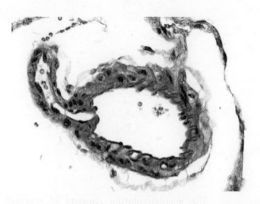

Figure 4.17 Thickening of the media at the division of a meningeal artery. Anatomical variations and oblique sections of blood vessels can be mistaken for medial hypertrophy.

arterioles being <100 μm (Gallagher and van der Wal 2007) (Figure 4.19).

On the venous side of the circulation the pressure is lower and vessels are therefore thinner walled and tend to be of larger diameter. Muscular venules and small veins are lined by endothelium and have a thin intima with no elastic tissue, a thin media with one or two muscle cell layers and the adventitia blends with surrounding connective tissue. Large muscular veins have no distinct internal elastic lamina, a muscular wall and broad adventitia. Valves occur in the largest veins, for example in the iliac veins, and are thin bicuspid structures lined by endothelium (Bazigou *et al.* 2011). In mice, the walls of pulmonary veins down to 70 μm in

diameter may contain variable amount of cardiac myofibres (Mueller-Hoecker *et al.* 2008).

The lymphatic vessels are often inconspicuous in tissue sections but they are present and mirror blood vessels in all organs except in the central nervous system, bone, teeth and cornea. If present they can be distinguished from capillaries by an absence of red blood cells in the lumen. Lymph vessels are lined by endothelium, have an incomplete

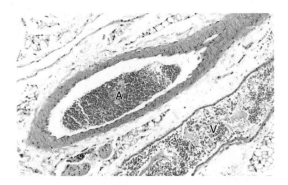

Figure 4.18 Comparison of a vein (V) and artery (A) in the mesentery of a mouse. The artery has a thicker medial layer, internal and external elastic laminae and an obvious adventitial layer. The vein has no obvious elastic lamina, a thin muscular media and indistinct adventitia.

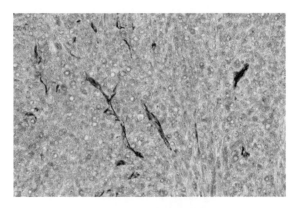

Figure 4.20 Immunohistochemical staining with an antibody against CD31 (PECAM1) demonstrating tumour blood vessels.

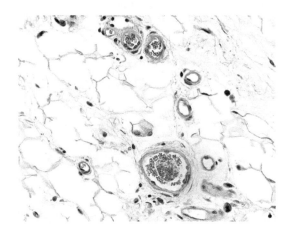

Figure 4.19 The microcirculation consists of precapillary arterioles, the capillary bed, collecting venules and postcapillary venules.

basement membrane and no pericytes. They are anchored to the surrounding connective tissue by thin collagen and elastic fibres to prevent collapse. Larger lymphatics have some smooth muscle in their walls and valves similar to those found in veins.

While there are a number of immunohistochemical markers that can be used to identify endothelium, due to the heterogeneous nature of endothelia in the micro and microvasculature and in different organs, no one marker is universally applicable (Kasper 2005). Conversely the markers that exist do not specifically label the different endothelial populations or help to distinguish lymphatic from vascular endothelium. Antibodies against CD31 and CD34 (Figure 4.20) are amongst the most commonly used for detecting vascular endothelium and LYVE1 and PROX-1 are used for lymphatic endothelium in mouse tissue (Ezaki *et al.* 2006).

References

Bailey, S.A., Zidell, R.H. and Perry, R.W. (2004) Relationships between organ weight and body/brain weight in the rat: what Is the best analytical endpoint? *Toxicologic Pathology* **32**, 448–466.

Bazigou, E., Lyons, O.T.A., Smith, A. *et al.* (2011) Genes regulating lymphangiogenesis control venous valve formation and maintenance in mice. *Clinical Investigation* **121** (8), 2984–2992.

Bishop, S.P. and Louden, C. (1997) Morphologic evaluation of the heart and blood vessels, in *Comprehensive Toxicology* (eds I.G. Sipes, C.A. McQueen and A.J. Gandolf), Elsevier Science Ltd, New York, NY, pp. 73–93.

Christoffels, V.M. and Moorman, A.F.M. (2009) Development of the cardiac conduction system: why are some regions of the heart more arrhythmogenic than others? *Circulation: Arrhythmia and Electrophysiology* **2**, 195–207.

Elwell M.R. and Mahler, J.F. (1999) Heart, blood vessels and lymphatic vessels, in *Pathology of the Mouse* (ed. R.R. Maronpot), Cache River Press, Saint Louis, MO, p. 362.

Ezaki, T., Kuwahara, K., Morikawa, S. *et al.* (2006) Production of two novel monoclonal antibodies that distinguish mouse lymphatic and blood vascular endothelial cells. *Anatomy and Embryology* **211**, 379–393.

Fernández, B., Durán, A.C., Fernández, M.C. *et al.* (2008) The coronary arteries of the C57BL/6 mouse strains: implications for comparison with mutant models. *Journal of Anatomy* **212**, 12–18.

Gallagher, P.J. and van der Wal, A.C. (2007) Blood vessels, in *Histology for Pathologists* (ed. S.E. Mills), 3rd edn, Lippincott, Williams & Wilkins, Philadelphia, PA, pp. 224–225.

Hall, A.P. (2006) Review of the pericyte during angiogenesis and its role in cancer and diabetic retinopathy. *Toxicologic Pathology* **34**, 763–775.

Hinton, R.B., Alfieri, C.M., Witt, S.A. *et al.* (2008) Mouse heart valve structure and function: echocardiographic and morphometric analyses from the fetus through the aged adult. *American Journal of Physiology. Heart and Circulatory Physiology* **294**, H2480–H2488.

Hort, W. (1971) Quantitative morphology and structural dynamics of the myocardium. *Methods and Achievments in Experimental Pathology* **5**, 3–21.

Frith, C. and Ward, J.M. (1988) Color Atlas of Neoplastic and Non-neoplastic Lesions in Aging Mice, www.informatics.jax.org/frithbook/frames/framecardio.shtml (accessed 16 July 2013).

Isaacs, K.R. (1998) The cardiovascular system, in *Target Organ Pathology: A Basic Text* (eds J.A. Turton and J. Hooson), Taylor & Francis, London.

Kasper, M. (2005) Phenotypic characterization of pulmonary arteries in normal and diseased lung. *Chest* **128**, 547S–552S.

Kaufman, M.H. and Bard, J.B.L. (1999) *The Anatomical Basis of Mouse Development*, Academic Press, San Diego, CA.

Keenan, C.M. and Vidal, J. (2006) Standard morphologic evaluation of the heart in the laboratory dog and monkey. *Toxicologic Pathology* **34**, 67–74.

Krinke, G.J. Normative histology of organs. Organ weights, in *The Laboratory Mouse* (ed. H. Hedrich), Elsevier Academic Press, London, p. 134.

Li, J. and Radice, G.L. (2010) A new perspective on intercalated disc organization: implications for heart disease. *Dermatology Research and Practice*, doi:10.1155/2010/207835.

Liu, J. and Rigel, D.F. (2009) Echocardiographic examination in rats and mice. *Methods in Molecular Biology* **573**, 139–155.

Liu, Z, Yue, S., Chen, X. *et al.* (2010) Regulation of cardiomyocyte polyploidy and multinucleation by CyclinG1 .*Circulation Research* **106**, 1498-1506.

Martinez, H.G., Prajapati, S.I., Estrada, C.A. *et al.* (2009) Microscopic computed tomography-based virtual histology for visualization and morphometry of atherosclerosis in diabetic apolipoprotein E mutant mice MD. *Circulation* **120**, 821–822.

Morawietz, G., Ruehl-Fehlert, C., Kittel, B. *et al.* (2004) Revised guides for organ sampling and trimming in rats and mice – Part 3. *Experimental Toxicologic Pathology* **55**, 433–449.

Mueller-Hoecker, J., Beitinger, F., Fernandez, B. *et al.* (2008) Of rodents and humans: a light microscopic and ultrastructural study on cardiomyocytes in pulmonary veins. *International Journal of Medical Science* **5**,152–158.

Olmesdahl, P.J., Gregory, M.A., and Cameron, E.W.J .(1979) Ultrastructural artefacts in biopsied normal myocardium and their relevance to myocardial biopsy in man. *Thorax* **34**, 82–90.

Qiao, J.-H., Fishbein, M.C., Demer, L.L. and Lusis, A.J. (1995) Genetic determination of cartilaginous metaplasia in mouse aorta. *Arteriosclerosis, Thrombosis, and Vascular Biology* **15**, 2265–2272.

Rentschler, S., Vaidya, D.M., Tamaddon, H. *et al.* (2001) Visualization and functional characterization of the developing murine cardiac conduction system. *Development* **128**, 1785–1792.

Roth, D.M., Swaney, J.S., Dalton, N.D. *et al.* (2002) Impact of anesthesia on cardiac function during echocardiography in mice. *American Journal of Physiology. Heart and Circulatory Physiology* **282**, H2134–H2140.

Savolainen, S.M., Foley, J.F. and Elmore, S.A. (2009) Histology atlas of the developing mouse heart with emphasis on E11.5 to E18.5. *Toxicologic Pathology* **37**, 395–414.

Sugiyama, F., Churchill, G.A., Li, R. *et al.* (2002) QTL associated with blood pressure, heart rate, and heart weight in CBA/CaJ and BALB/cJ mice. *Physiological Genomics* **10**, 5–12.

Taylor, I. (2011) Mouse, in *Background Lesions in Laboratory Animals* (ed. E.F. McInnes), Saunders, Elsevier, Edinburgh, p. 44.

Wessels, A. and Sedmera, D. (2003) Developmental anatomy of the heart: a tale of mice and man. *Physiological Genomics* **15** (3), 165–176.

Chapter 5
Urinary system

Elizabeth McInnes
Cerberus Sciences, Thebarton, Australia

5.1   Background and development

The urinary system of the mouse comprises the paired kidneys, urinary bladder, paired ureters and urethra. The murine kidney demonstrates similar structures to those of other mammalian species and many mouse models of human kidney disease have been developed by researchers.

Mesodermal cells immediately adjacent to the intermediate mesoderm form the urogenital system, comprising the gonads, the sex ducts and the kidneys (Kispert and Gossler 2004). The development of the mouse kidney is characterised by the formation of three paired embryonic kidney structures called the pronephros, mesonephros and metanephros. These three paired kidney structures develop from mesodermal cells. The pronephros and mesonephros degenerate leaving the metanephros which develops into the adult kidney. The definitive adult kidney develops through a series of inductive interactions between the uteric bud and the surrounding metanephric mesenchyme (Ohse *et al.* 2009). The tubules of the mesonephros do not form part of the permanent kidney. The Wolffian duct (also known as mesonephric duct) is a paired organ found in mammals during embryogenesis. It connects the primitive kidney Wolffian body (or mesonephros) to the cloaca and serves as the foundation for certain male reproductive organs. The ureter buds from the Wolffian duct and the urinary bladder is formed partly from the endodermal cloaca and the Wolffian ducts.

The metanephric mesenchyme develops into the renal vesicle, which forms the functional kidney. Collecting tubules (Figure 5.1) and primitive glomeruli can be recognized from about embryonic day 14, at which stage the kidney starts to produce urine. Well differentiated cortex and medulla containing glomeruli and nephrons can be distinguished from about embryonic Day 16 (Kaufman and Bard 1999).

A Practical Guide to the Histology of the Mouse, First Edition. Cheryl L. Scudamore.
© 2014 John Wiley & Sons, Ltd. Illustrations © Veterinary Path Illustrations, unless stated otherwise.
Published 2014 by John Wiley & Sons, Ltd. Companion Website: www.wiley.com/go/scudamore/mousehistology

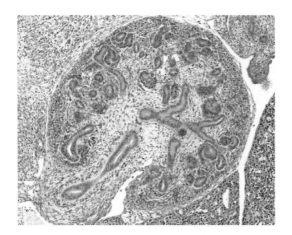

Figure 5.1 Metanephric development at embryonic day E14.5, with the start of differentiation of nephrons and the collecting duct system.

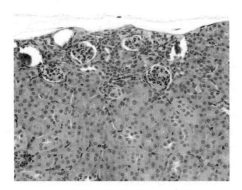

Figure 5.2 Occasional immature glomeruli may be retained even in mature animals.

The kidney continues to mature anatomically and functionally over the first five to six weeks of life in the mouse. Regression of glomeruli occurs over the first two to three weeks of life (Zhong *et al.* 2012) but occasional immature glomeruli may be retained even in mature animals (Figure 5.2).

5.2   Sampling techniques

Both kidneys should be harvested at necropsy and it may be useful to make a longitudinal incision in the left kidney and a transverse incision in the right kidney at necropsy if it is important to distinguish left and right kidneys from one another. Both kidneys should be fixed for 24 hours (minimum) in 10% neutral buffered formalin before cutting either longitudinal or transverse sections of kidney and placing the tissue in a cassette. At necropsy it should be possible to visualize the outer zone of the kidney, the cortex, as well as the inner zone of the kidney called the medulla and the pelvis, which is the structure created from the expanded upper portion of the ureter (Pakurar and Bigbee 2009). The easiest way to ensure that all zones of the kidney are consistently sampled is by taking transverse sections. Longitudinal sections may be justified if there is suspicion that a lesion may only occur at the poles – which is unusual but may occur if a tumour is present.

The ureters are harvested by gently lifting the adipose tissue between the kidneys and the bladder and exposing the left and right ureters by removing the surrounding adipose tissue. The bladder is located at the base of the uterus (in female mice) and above the prostate gland in male mice. The bladder should be sectioned as close to the underlying organs (uterus in female mice and prostate in male mice) as possible. Infiltration of the bladder with formalin may be necessary in order to remove folds in the inner urothelium.

5.3   Artefacts

Artefacts in the kidney include the extrusion of renal cortical epithelial cells into the Bowman's space due to excessive handling of the kidney at necropsy. Kidneys are also particularly susceptible to autolysis and this can be difficult to distinguish from acute tubular necrosis. Autolysed renal tissue is generally characterised by the presence of pale eosinophilic material and cuboidal epithelium within the cortical tubules, particularly epithelial cells without distinct borders and indistinct nuclei (Figure 5.3). Kocovski and Duflou (2009) reported that autolysis is extremely difficult to distinguish from acute tubular necrosis in the kidney and that only the presence of epithelial whorls within the tubules is always characteristic of acute tubular necrosis. Periodic Acid-Schiff (PAS)

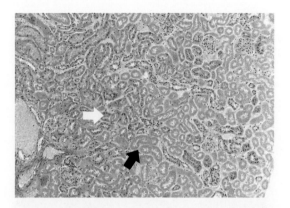

Figure 5.3 Autolysed renal tissue is generally characterised by the presence of pale eosinophilic material and cuboidal epithelium within the cortical tubules, epithelial cells without distinct borders or obvious nuclei (black arrow) or pyknotic nuclei (white arrow).

staining may be helpful in identifying autolysis in the kidney as a reduction in PAS positive staining has been reported to correlate with the onset of autolysis (Zdravkovic *et al.* 2006).

5.4   Background lesions

Background lesions are seen in urinary tract of young and old mice and have been reviewed recently (Taylor 2011 and Frazier *et al.* 2012). Familiarity with these lesions is important so that they can be differentiated from induced lesions. The investigator should also be aware that changes in incidence or severity of these lesions may be induced and therefore may represent a genuine phenotype.

Hydronephrosis (pelvic dilatation) (Goto *et al.* 1984) (Figure 5.4) and cysts (Figure 5.5) within the cortex or medulla of the kidney are common background lesions in the mouse kidney. Hydronephrosis may be congenital or acquired (in the case of obstruction in the ureters, bladder or urethra) and may be unilateral or bilateral.

Chronic progressive nephropathy (CPN) is observed commonly in ageing mice (Figure 5.6) and is characterised by thickening of the glomerular basement membranes, obliteration of the glomeruli,

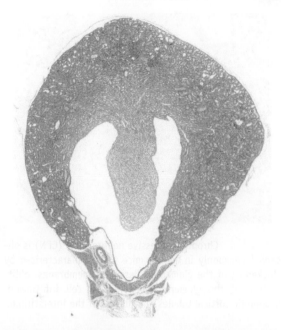

Figure 5.4 Hydronephrosis or pelvic dilatation may be congenital or acquired (in the case of obstruction in the ureters, bladder or urethra) and may be unilateral or bilateral.

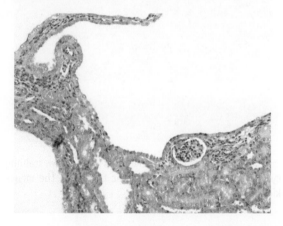

Figure 5.5 Cysts within the cortex or medulla of the kidney are common background lesions in the mouse kidney.

mononuclear cell infiltration, basophilic cortical tubules and fibrosis of the interstitium (Percy and Barthold 2007). Even in mice as young as five weeks, the early changes of chronic progressive nephropathy may be present and these include lymphocyte aggregates in the interstitial cortex and cortical

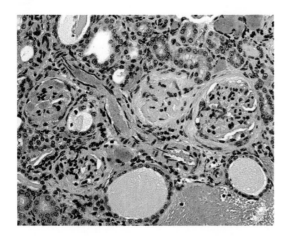

Figure 5.6 Chronic progressive nephropathy (CPN) is observed commonly in ageing mice and is characterised by thickening of the glomerular basement membranes, obliteration of the glomeruli, mononuclear cell infiltration, basophilic cortical tubules and fibrosis of the interstitium.

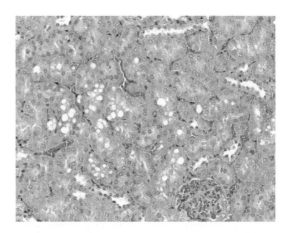

Figure 5.8 Vacuolation is a common background finding in cortical epithelium.

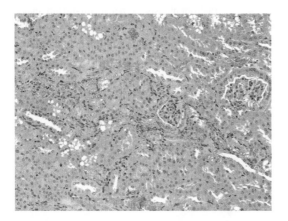

Figure 5.7 Cortical epithelial cells with blue staining cytoplasm, referred to as basophilic tubules in the mouse kidney.

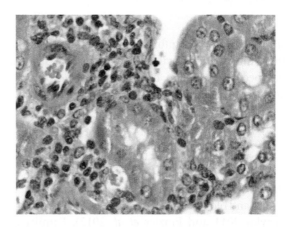

Figure 5.9 Interstitial inflammation (particularly lymphocytes) in the cortex of the mouse kidney.

epithelial cells with blue staining cytoplasm referred to as basophilic tubules (Figure 5.7).

Other changes that are commonly seen in tubules are vacuolation (Figure 5.8) and pigmentation as a result of accumulations of lipid and lipofucsin respectively (Frazier *et al.* 2012).

Interstitial changes include adipose aggregates (lipomatosis, lipomatous metaplasia), inflammation (particularly lymphocytes) (Figure 5.9) and mineralization (Figure 5.10), which may occur in the cortical, medullary or papillary regions of the mouse kidney (Seely 1999).

Refluxed seminal colloid plugs in the bladder (Figure 5.11) and urethral plugs (copulatory plugs) occur occasionally in mice. Ejaculation results in the formation of a copulatory plug and this often occurs agonally after euthanasia (Percy and Barthold 2007). The yellow plug is composed of eosinophilic, proteinaceous material within which spermatozoa may be embedded (Taylor 2011). Urolithiasis or the formation of urinary stones may be distinguished

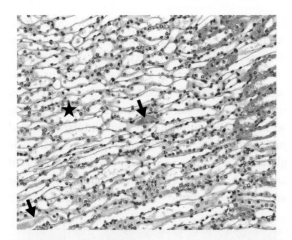

Figure 5.10 The collecting tubule drains urine from the distal convoluted tubules in the cortex and descends into the medulla where it joins other collecting tubules (*) in the papilla. Small foci of mineralization are commonly found in the papilla (arrows) as well as other regions of the kidney.

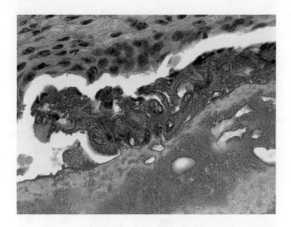

Figure 5.11 Refluxed seminal colloid plugs in the bladder are composed of eosinophilic, proteinaceous material within which spermatozoa may be embedded.

from colloidal reflux by the presence of inflammatory cells in the lumen and the adjacent epithelium and the mineralized concentric layered appearance of genuine uroliths.

Obstructive uropathy (urologic syndrome) is observed commonly in male mice on long-term studies (Maita *et al.* 1988) and may be a cause of death. This change is associated with group housing of male mice and is related to fighting and genital wounds. At necropsy one may observe marked distension and congestion (reddening) of the urinary bladder as well as a swollen and enlarged penis and haemorrhage around the bulbourethral gland area.

5.5 Anatomy and histology

The urinary system is made up of the paired kidneys and ureters as well as the urinary bladder and urethra. Urine is delivered from the kidneys to the urinary bladder by the ureters and the urinary bladder is emptied via the urethra.

5.5.1 *Kidneys*

The kidneys are reddish, bean-shaped organs situated on the dorsal abdominal wall. In the mouse, the right kidney is more cranial than the left kidney and the adrenal glands and adipose tissue are present at the cranial poles of the kidneys. The male mouse kidneys are larger than those present in female mice and there are strain variations size (Murawski *et al.* 2010).

The mouse kidney has a single wedge-shaped papilla (Krinke 2004) projecting into the renal sinus and surrounded by the renal pelvis, which is continuous with the ureter (Frazier *et al.* 2012). Blood vessels, lymphatics and the ureters enter/exit the kidney at the hilus. The kidney is covered by a loose capsule, which may be removed using forceps. Underlying pathology can make the capsule more adherent and difficult to remove but, in practice, for routine histology in mice, the capsule should be left attached to the underlying parenchyma to reduce the risk of damage to the underlying tissue.

The kidney is anatomically divided macroscopically into the cortex and medulla. The mouse kidney papilla extends quite far forward into the pelvis and ureter compared to other domestic animals (Figure 5.12). The mouse kidney can be further divided into five topographic areas, which can be distinguished histologically and these include the cortex, the outer stripe of the outer medulla, the

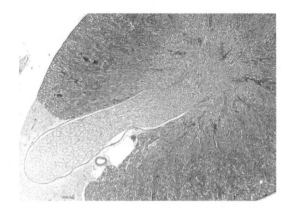

Figure 5.12   The mouse kidney papilla extends quite far forward into the pelvis and ureter compared to other domestic animals.

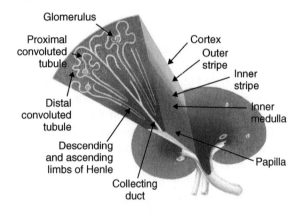

Figure 5.13   The mouse kidney can be divided into five topographic areas, which can be distinguished histologically and these include the cortex, the outer stripe of the outer medulla, the inner stripe of the outer medulla, the inner medulla and papilla. The nephron is made up of the glomerulus, proximal convoluted tubule, descending and ascending limbs of Henle, distal convoluted tubule and collecting duct system.

inner stripe of the outer medulla, the inner medulla and papilla (Frazier *et al.* 2012) (Figure 5.13). The renal artery enters the kidney and gives rise to the interlobular, arcuate and segmental arteries (Treuting and Kowalewska 2012).

The kidneys are made up of nephrons and collecting ducts (Figure 5.13). There are approximately 20 000 nephrons in the adult mouse kidney, but there is some variation by strain and age, with progressive

loss over time (Murawski *et al.* 2010,). Mice have relatively large numbers of glomeruli per unit area compared with the rat (Percy and Barthold 2007).

The nephron is made up of the glomerulus, proximal convoluted tubule, descending and ascending limbs of Henle, distal convoluted tubule, connecting segment, collecting duct system, interstitium and juxtaglomerular apparatus (Figure 5.14). The glomerulus (Figure 5.15) is supplied by the afferent arteriole and is emptied by the efferent glomerular arteriole. Glomeruli (or renal corpuscles) are

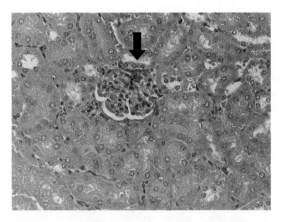

Figure 5.14   The juxtaglomerular apparatus (arrow) is situated at one end of the glomerulus and is made up of the macula densa, the efferent and afferent arterioles and the renin-secreting granular cells of the afferent arteriole.

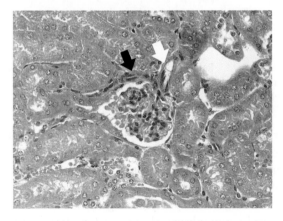

Figure 5.15   The glomerulus is supplied by the afferent arteriole (black arrow) and is emptied by the efferent glomerular arteriole (white arrow).

concentrated in the cortex. The glomerulus is made up of a nest of capillaries, which project into the Bowman's space. The Bowman's capsule surrounds the urinary space adjacent to the glomerulus and this is lined by parietal epithelial cells (Ohse *et al.* 2009). Parietal epithelial cells may transform into podocytes and play a role in limiting filtered albumin from leaving the urinary space (Ohse *et al.* 2009). Parietal epithelial cells demonstrate a high percentage of proliferating cells (Cheval *et al.* 2011). The glomerular tuft is lined by endothelial cells that rest on a basal lamina with openings, opposite a layer of podocytes. The central adjacent region of the glomerulus is made up of myoepithelial cells called mesangial cells and there is a single glomerular basement membrane between endothelial cells and podocytes. Glomerular size tends to increase in ageing mice and is strain related (Frazier *et al.* 2012). Special stains including PAS and silver stains may be used to visualize the glomerular basement membrane (Figure 5.16) (Treuting and Kowalewska 2012).

In male mice, the proximal convoluted tubule cells project into the Bowman's capsule. Mice display sexual dimorphism in the parietal epithelium of Bowman's capsule. This means that in mature male mice, the parietal epithelium of Bowman's capsule is generally cuboidal (Figure 5.17) whereas in female mice, the epithelium is generally flattened (Figure 5.18) (Yabuki *et al.* 2003). In addition, the

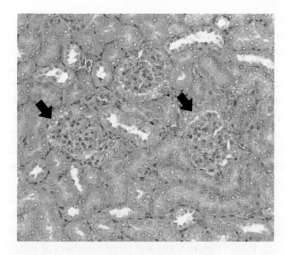

Figure 5.17　Mice display sexual dimorphism in the parietal epithelium of Bowman's capsule. This means that in mature male mice, the parietal epithelium of Bowman's capsule is generally cuboidal (arrow).

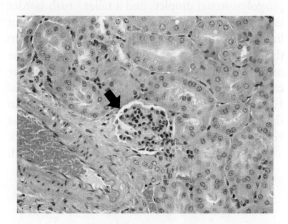

Figure 5.18　In female mice, the parietal epithelium of the Bowman's capsule is generally flattened (arrow).

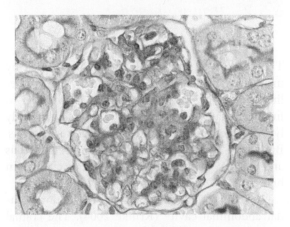

Figure 5.16　Special stains including PAS may be used to visualize the glomerular and tubular basement membranes.

percentage of renal corpuscles with cuboidal cells and the diameter of renal corpuscles is larger in male than female mice (Yabuki *et al.* 1999). Large numbers of immature glomeruli are visible in the cortex in young mice; however these disappear as the mouse assumes maturity (Figure 5.2).

The juxtaglomerular apparatus is situated at one end of the glomerulus and is made up of the macula densa, the efferent and afferent arterioles and the renin-secreting granular cells of the afferent arteriole (Figure 5.14). Granular cells are modified pericytes of glomerular arterioles. Macula densa cells are columnar epithelial cells of the distal tubule. The smooth muscle cells of the afferent arteriole produce renin and are responsible (together with the macula densa and the extraglomerular mesangial cells) for controlling the animal's blood pressure (Treuting and Kowalewska 2012). Immunohistochemistry (anti-renin antibodies), PAS special staining and electron microscopy may be used to distinguish the granular cells.

The proximal convoluted tubules are divided into S1, S2 and S3 segments. The S1 segment in the mouse has a well developed brush border (positive with PAS) (Figure 5.19) and large numbers of mitochondria and has a circular lumen in cross section. The S2 segment has a shorter brush border, less mitochondria and often has tubular convolutions in cross section. The S3 or straight segment has sparse phagolysosomal droplets and a taller brush border than other segments (Frazier *et al.* 2012). The thin limb of the descending loop of Henle exits from the S3 segment, forming the boundary between the outer and inner stripes of the outer medulla (Frazier *et al.* 2012). The thin ascending limb of the loop of Henle becomes the thick ascending loop, returns to the cortex within a medullary ray and terminates just beyond its own glomerulus. Cells lining the thick ascending limb are cuboidal and eosinophilic, with prominent, oval, central nuclei and these epithelial cells are smaller than the cells found in the proximal convoluted tubules. Immunohistochemistry may be used to identify different segments of the proximal convoluted tubule and distal convoluted tubule in the mouse (Bauchet *et al.* 2011). Aquaporin-1 is specific for the proximal tubules and the thin descending limbs of Henle's loops in mice (Bauchet *et al.* 2011). In addition, vacuolar structures in the proximal convoluted tubules of male DBA/2 mice have been shown to be giant lysosomes containing apolipoprotein (Yabuki *et al.* 2002). Tamm-Horsfall protein can be used to stain the thick ascending limbs of Henle's loops in mice (Bauchet *et al.* 2011). Calbindin-D$_{28K}$ may be used to stain the distal convoluted tubules and cortical connecting and collecting ducts of mice (Bauchet *et al.* 2011).

The collecting ducts extend from the cortex, through the medullary ray and inner and outer medulla to the tip of the papilla and are lined with predominantly low columnar epithelium. The apex of the renal pyramid is the renal papilla and this opens into the calyx, which, in turn, empties into the renal pelvis (Pakurar and Bigbee 2009). The collecting tubule (Figure 5.10) drains urine from the distal convoluted tubules in the cortex and descends into the medulla where it joins other collecting tubules to form a papillary ducts and aids in concentrating the urine.

5.5.2 *Urinary bladder, ureter and urethra*

The mouse renal pelvis (Figure 5.20) is lined by urothelium similar, but thinner than the epithelium lining the urinary bladder. It is common to find aggregates of inflammatory cells, particularly lymphocytes in the submucosa of the renal pelvis. Brown fat may be noted around the renal pelvis and this may be a useful area to evaluate changes in brown fat.

The urinary bladder is located in the dorsocaudal abdominal cavity (Krinke 2004). The cranial portion of the mouse urinary bladder is called the dome and the caudal area is called the trigone. The

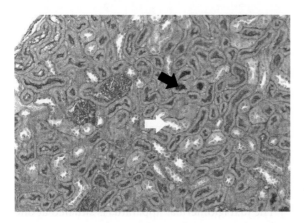

Figure 5.19 PAS also highlights the presence of a brush border in the proximal convoluted tubule (black arrow) and its absence in the distal convoluted tubule (white arrow).

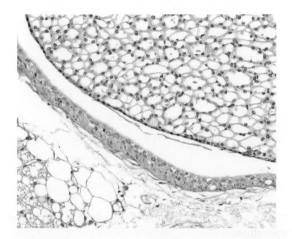

Figure 5.20 The mouse renal pelvis is lined by urothelium similar, but thinner than the epithelium lining the urinary bladder.

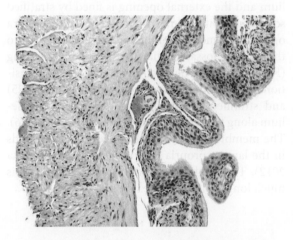

Figure 5.21 The urothelium of the urinary bladder, ureter and urethra is a multilayered epithelium and the surface is lined by umbrella cells, which are large and polygonal and may be binucleate.

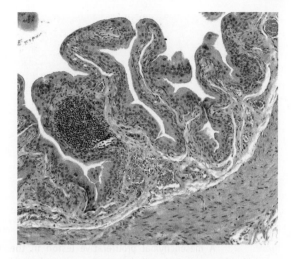

Figure 5.22 In the mouse, multifocal lymphocyte or mast cell aggregates are noted commonly in the submucosa of the bladder.

urothelium (Figure 5.21) of the urinary bladder, ureter and urethra is a multilayered epithelium and the surface is lined by umbrella cells, which are large and polygonal and may be binucleate (Treuting and Kowalewska 2012). The urothelium is surrounded by a layer of thick connective tissue, two layers of smooth muscle – inner circular and outer longitudinal and an outer adventitia. In the mouse, multifocal lymphocyte or mast cell aggregates are

noted commonly in the submucosa of the bladder (Figure 5.22). The mucosa of the bladder is arranged into numerous folds that disappear when the bladder is distended with urine. The prosector at necropsy may wish to inflate the bladder with formalin using a syringe and needle to avoid the formation of artifactual folds and thickening in the bladder wall. The lamina propria of the bladder does not contain glands except in the region surrounding the exit of the urethra where mucous glands are situated.

A mesenchymal proliferative lesion in the urinary bladder may occasionally be seen in ageing mice and consists of spindle and epithelioid cells, occasional round, eosinophilic granules and may possess nuclear progesterone receptors and cytoplasmic desmin (Karbe *et al.* 1998) (Figure 5.23). This lesion appears to develop from mesenchymal cells near the trigone area of the urinary bladder (Karbe *et al.* 1998).

The ureters are muscular tubes that collect the urine and carry it to the urinary bladder. The ureters of the mouse kidney are extremely thin. They may be identified by lifting the retroperitoneal adipose tissue between the kidney and the bladder. The ureters are made up of urothelium, lamina propria smooth muscle bundles and an outer adventitia (Figure 5.24).

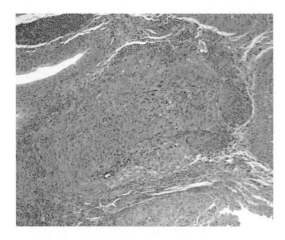

Figure 5.23   A mesenchymal proliferative lesion in the urinary bladder may occasionally be seen in ageing mice and consists of spindle and epithelioid cells, occasional round, eosinophilic granules and may possess nuclear progesterone receptors and cytoplasmic desmin.

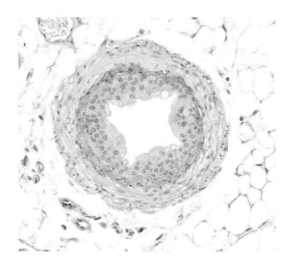

Figure 5.24   The ureters are made up of urothelium, lamina propria smooth muscle bundles and an outer adventitia.

The urethra is divided in to the membranous portion, which extends from the bladder to the penis and the penile portion which extends along the length of the penis (Treuting and Kowalewska 2012). Before its transition to the penile portion, the urethra forms a diverticulum and receives the

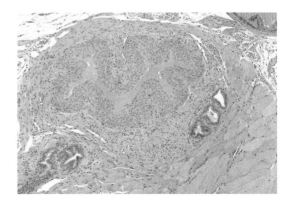

Figure 5.25   Urethra near the bladder lined by transitional epithelium.

openings of the bulbourethral glands (Krinke 2004). The lumen of the penile urethra is lined by urothelium and the external opening is lined by stratified squamous epithelium. In female mice, the urethra opens independently of the vagina and empties into the clitoral fossa, cranial to the vaginal opening (Krinke 2004). The urethra is lined with transitional epithelium near the bladder (Figure 5.25) and stratified squamous, non-keratinized epithelium along the remainder of its length (Figure 5.26). The membranous urethra has many mucus glands in the lamina propria (Treuting and Kowaleswska 2012). The urethra is short in female mice and is much longer in male mice.

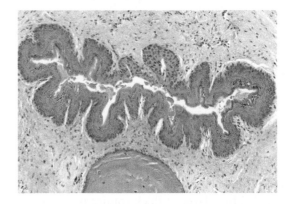

Figure 5.26   Penile urethra lined by stratified squamous epithelium.

References

Bauchet, A.L., Masson, R., Guffroy, M. and Slaoui, M. (2011) Immunohistochemical identification of kidney nephron segments in the dog, rat, mouse, and cynomolgus monkey. *Toxicologic Pathology* **39**, 1115–1128.

Cheval, L., Pierrat, F., Dossat, C. *et al.* (2011) Atlas of gene expression in the mouse kidney: new features of glomerular parietal cells. *Physiological Genomics* **43**, 161–173.

Frazier, K.S., Seely, J.C., Hard, G.C. *et al.* (2012) Proliferative and nonproliferative lesions of the rat and mouse urinary system. *Toxicologic Pathology* **40** (4 suppl.), 14S–86S.

Goto, N., Nakajima, Y., Onodera, T. *et al.* (1984) Inheritance of hydronephrosis in the inbred mouse strain DDD. *Laboratory Animal* **18**, 22–25.

Karbe, E., Hartmann, E., George, C. *et al.* (1998) Similarities between the uterine decidual reaction and the 'mesenchymal lesion' of the urinary bladder in aging mice. *Experimental Toxicologic Pathology* **50**, 330–340.

Kaufman, M. and Bard, J. (1999) *The Anatomical Basis of Mouse Development*, Academic Press, San Diego, CA, pp. 109–116.

Kispert, A. and Gossler, A. (2004) Introduction to early mouse development, in *The Laboratory Mouse* (eds H. Hedrich and G. Bullock), Elsevier, New York, pp. 175–181.

Kocovski, L. and Duflou, J. (2009) Can renal acute tubular necrosis be differentiated from autolysis at autopsy? *Journal of Forensic Sciences* **54**, 439–442.

Krinke, G.J. (2004) Normative histology, in *The Laboratory Mouse* (eds H. Hedrich and G. Bullock), Elsevier, New York, pp. 144–145.

Maita, K., Hirano, M., Harada, T., *et al.* (1988) Mortality, major cause of moribundity, and spontaneous tumors in CD-1 mice. *Toxicologic Pathology* **16**, 340–349.

Murawski, I.J., Maina, R.W., Gupta, I.R. (2010) The relationship between nephron numbers, kidney size and body weight in two inbred mouse strains. *Organogenesis* **6**, 189–194.

Ohse, T., Pippin, J.W., Chang, A.M. *et al.* (2009) The enigmatic parietal epithelial cell is finally getting noticed: a review. *Kidney International* **76**, 1225–1238.

Pakurar, A.S. and Bigbee, J.W. (2009) Urinary system, in *Digital Histology*, 2nd edn (eds A.S. Pakurar and J.W. Bigbee), Wiley-Blackwell, Hoboken, NJ, pp. 159–167.

Percy, D.H. and Barthold, S.W. (2007) *Pathology of Laboratory Rodents and Rabbits*, 3rd edn, Blackwell, Ames, IA, pp. 3–124.

Seely, J.C. (1999) Kidney, in *Pathology of the Mouse, Reference and Atlas* (eds R.R. Maronpot, G.A. Boorman and B.W. Gaul), Cache River Press, Vienna, pp. 207–234.

Taylor, I. (2011) Mouse, in *Background Lesions in Laboratory Animals* (ed. E.F. McInnes), Saunders, Edinburgh, pp. 45–72.

Treuting, P.M. and Kowalewska, J. (2012) Urinary system, in *Comparative Anatomy and Histology: A Mouse and Human Atlas* (eds P.M. Treuting and S. Dintzis), Elsevier, Amsterdam, pp. 229–253.

Yabuki, A., Matsumoto, M., Nishinakagawa, H. and Suzuki, S. (2003) Age-related morphological changes in kidneys of SPF C57BL/6Cr mice maintained under controlled conditions. *Journal of Veterinary Medical Science* **65**, 845–851.

Yabuki, A., Suzuki, S., Matsumoto, M. and Nishinakagawa, H. (1999) Morphometrical analysis of sex and strain differences in the mouse nephron. *Journal of Veterinary Medical Science* **61**, 891–896.

Yabuki, A., Suzuki, S., Matsumoto, M. and Nishinakagawa, H. (2002) Sex- and strain-dependent histological features of the proximal convoluted tubular epithelium of mouse kidney: association with lysosomes containing apolipoprotein B. *Histology and Histopathology* **17**, 1–7.

Zdravkovic, M., Kostov, M. and Stojanovic, M. (2006) Identification of post-mortem autolytic changes on the kidney tissue using PAS stained method. *Facta Universitatis* **13**, 181–184.

Zhong, J., Perrien, D.S., Yang, H.C. *et al.* (2012) Maturational regression of glomeruli determines the nephron population in normal mice. *Pediatric Research* **72**, 241–248.

References

Chapter 6
Reproductive system

Cheryl L. Scudamore
Mary Lyon Centre, MRC Harwell,UK

6.1   Background and development of the male and female reproductive tract

The reproductive organs in both sexes are derived principally from embryonic mesoderm and the incorporation of primordial germ cells. There is a small contribution of ectoderm to the formation of the external genitalia. Primordial germ cells, which are the indifferent precursors of sperm and ova, migrate from the yolk sac to the genital ridges early in embryonic development. Occasional abnormal migration leads to abnormal deposition of these cells in extragondal tissues, which can lead to extragonadal teratoma formation in adult animals. Differentiation of the gonads towards the appropriate genders specific organ is a complex process but is generally believed to be initiated by the presence of sex determining factors such as Sry from the Y chromosome.

The tubular structures of the reproductive tract develop from the mesonephric or Wolffian duct and the paramesonephric or Mullerian duct. In the presence of developing testes the production of testosterone and Mullerian inhibiting substance leads to the development of the Wolffian duct system into the seminal vesicles, epididymides and vas deferens from the urogenital sinus and the regression of the Mullerian duct. In the absence of testosterone the Wolffian ducts regress and the Mullerian duct system develops into the uterus, oviduct and vagina (Pritchett and Taft 2007). The caudal genital tract (prostate and ampullary glands) and external genitalia develop in both sexes from the urogenital sinus. At birth the genital tubercle and the anogenital distance are larger in males than females but the external genitalia are still relatively undifferentiated and do not complete development until day 10. The penis and the clitoris derive from an ambivalent genital tubercle with sex differentiation occurring from day 16 of gestation.

Puberty occurs at four to six weeks depending on the mouse strain (Nelson *et al.* 1990) and environmental cues and is defined by the onset of reproductive competence. In females the formation

A Practical Guide to the Histology of the Mouse, First Edition. Cheryl L. Scudamore.
© 2014 John Wiley & Sons, Ltd. Illustrations © Veterinary Path Illustrations, unless stated otherwise.
Published 2014 by John Wiley & Sons, Ltd. Companion Website: www.wiley.com/go/scudamore/mousehistology

of the vaginal opening is a sign of the onset of puberty and generally marks the onset of cyclicity, although animals may not be able to sustain a pregnancy immediately. In males puberty is defined by the production of mature sperm although the actual ability to fertilize females may only occur several weeks after mature sperm are first observed (Creasey 2011).

6.2 Female reproductive tract

6.2.1 *Sampling technique*

Routine analysis of the female reproductive tract should also include analysis of the stage of the oestrous cycle. Although the stage of the oestrous cycle is often assessed on vaginal epithelial morphology, via vaginal smears or histological analysis, it is important to confirm that the rest of the tract including the uterus and ovaries are in synchrony with the appearance of the vagina. For routine analysis immersion in 10% NBF or paraformaldehyde is adequate for fixation of all elements of the tract. *In vivo*, the stage of the oestrous cycle can be assessed by evaluation of the external genitalia and taking vaginal smears. The vaginal opening varies from a tightly closed slit during dioestrus to being open and surrounded by swollen tissue at proestrus (Champlin *et al*. 1973; Byers *et al*. 2012). The stage can be further confirmed by vaginal cytology following the preparation of smears (Byers *et al*. 2012).

Normal mouse ovaries are very small so to ensure that they are not lost during tissue processing and to aid orientation for sectioning it is best to prepare them in a separate wax block from the rest of the reproductive tract. Biopsy pads or cassette inserts can be used to prevent loss of tissue during processing (Figure 6.1). If the ovary does not need to be weighed or inspected in detail at necropsy then it can be retained in its bursa with the associated oviduct. Longitudinal sections through the middle of the ovary will be sufficient for routine analysis and will also usually yield transverse and oblique

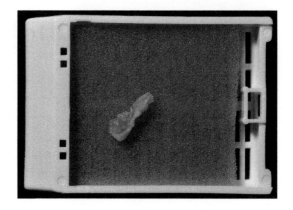

Figure 6.1 Ovary in a cassette with a biopsy pad.

sections of oviduct as well as the bursa. For longitudinal sections of oviduct the tip of the uterine horn and oviduct should be dissected and embedded separately (Kittel *et al*. 2004).

Microscopic analysis of the general populations of corpora lutea and follicles in longitudinal sections of ovary stained with haematoxylin and eosin, in conjunction with evaluation of the rest of the reproductive tract, are generally sufficient to assess whether an animal is cycling normally (Regan *et al*. 2005). However in some circumstances, for example in the evaluation of the effect of genetic or other manipulations on folliculogenesis, it may be necessary to quantify the number of follicles present within the ovary. In these circumstances a more systematic approach to sampling is required. A variety of stereological approaches based on the evaluation of multiple sections per ovary with correction factors are described and debated in the literature (Tilly 2003). In postnatal studies, ovaries should be harvested from animals at a defined stage in the oestrous cycle (Myers *et al*. 2004) and after routine processing serial sections made and the number of follicles of different types evaluated in a pre-determined subset of the slides.

The tubular reproductive organs can be prepared in a single wax block. For routine analysis a longitudinal section through the vagina, cervix and body of the uterus with transverse sections from each uterine horn (Kittel *et al*. 2004) is usually sufficient.

Rarely, it may be necessary to examine specifically the vulva and/or the clitoris and clitoral glands, which lie just cranial to the vulva in female mice. The area of the vulva including the glands can be dissected out and separated from the vagina and embedded to enable longitudinal sections through the area (Ruehl-Fehlert *et al.* 2003).

Examination of the placenta is not done routinely in phenotypic screening but may be required for specific investigations. Careful selection of appropriate samples is required to allow analysis of the pregnant uterus, placenta and foetal contents (see Blackburn 2000). Sections of uterus should be selected by cutting either side of the gestational sites and embedding the tissue so that longitudinal sections are taken. Serial sections can be made through the site or a single section at the maximum diameter taken for comparison between gestational sites or animals.

6.2.2   *Anatomy and histology*

The general anatomical structure of the mouse female reproductive tract is similar to most mammals that have multiple offspring and consists of paired ovaries, a duplex uterus with a relatively short body and long uterine horns, cervix and vagina. Significant variations in histological appearance occur with age and stage of the oestrous cycle and an awareness of these changes is important when analysing the tissues. In mature (greater than 5–6 months) fertile mice the oestrous cycle is generally around 4 days (Nelson *et al.* 1982) if the animal is not mated or pseudopregnant. Between puberty and full maturity the cycle length tends to be longer >5 days (Nelson *et al.* 1982). From about 12 months (Nelson *et al.* 1982) (although there is considerable variation depending on strain, nutrition and husbandry factors) the mouse starts to become reproductively senescent, with increases in cycle length and eventual loss of normal cyclicity.

The ovary is contained within a thin-walled bursa (Figure 6.2), which originates from mesovarium that, in turn, attaches the ovary to the peritoneum. The ovarian bursa is usually found embedded in fat

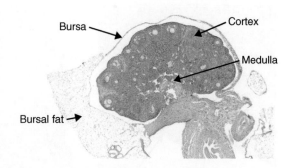

Figure 6.2   Overview of ovary and bursa.

just caudal to the kidneys. The bursa is lined on both sides by flattened mesothelial cells and has a thin connective tissue core which contains scattered smooth muscle fibres. Cystic distension of the bursa is common in older mice and needs to be distinguished from cystadenoma. Haemorrhagic ovarian cysts are also common in ageing mice and can become very large (>2 cm in diameter) (Chapter 1, Figure 1.37). Rupture can be associated with significant blood loss, anaemia and even sudden death.

The ovary has a surface epithelial layer composed of flattened to cuboidal epithelial cells (Figure 6.3) derived from the peritoneal lining cells which are attached to a thin basement membrane. The mouse ovary has an outer cortex that contains the follicular structures and corpora lutea and a central medulla which contains stroma and blood vessels (Figure 6.2). The division between cortex and medulla is ill defined and, because of the small size of the ovary, the medulla may not be present in every section. The ovarian stroma contains spindle shaped cells interspersed with collagen and supports the follicles and corpora lutea. The collagen is more dense in the area below the surface epithelium and the region is sometimes called the tunica albuginea although it is poorly defined and may be hard to distinguish in mice.

Ovarian follicles are categorized based on the morphological appearance of the associated granulosa cell layers into primordial, primary, secondary, antral and atretic follicles (Figures 6.3 and 6.4) (Myers *et al.* 2004). Primordial follicles tend to be found towards the subcapsular region of the ovarian

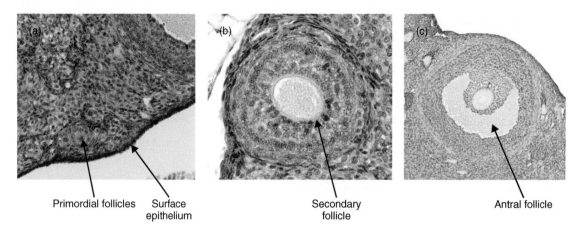

Primordial follicles   Surface
                        epithelium

Secondary
follicle

Antral follicle

Figure 6.3   Follicular types in the ovary. (a) Primordial follicles, (b) secondary follicle, (c) antral follicle.

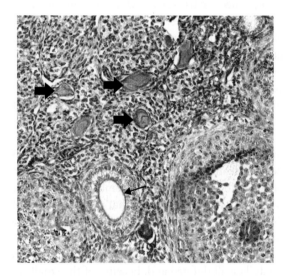

Figure 6.4   Atretic follicles (thick arrows) and zona pellucida (thin arrow) demonstrated with PAS stain.

cortex and consist of an oocyte surrounded by a single layer of flattened granulosa cells. Polyovular follicles may be seen in young mice but are rare in mature animals (Kent 1960). The growing oocytes are surrounded by a hyalinized layer of glycoproteins secreted by the oocyte (El-Mestrah *et al.* 2002) called the zona pellucida. In primary follicles the oocyte and zona pellucida is surrounded by a single layer of cuboidal granulosa cells and an outer layer

of flattened cells. Secondary follicles have multiple layers of cuboidal granulosa cells in close association with the oocyte, with no antral space. Antral follicles have multiple layers of granulosa cells and a clearly visible fluid-filled antral space (or spaces). The numbers of large antral follicles increase during proestrus and decrease as a result of ovulation during oestrus. In large preovulatory antral follicles, the oocyte is separated from a single antral space by a surrounding layer of granulosa cells (the cumulus granulosa), which will be retained with the oocyte when it is released at ovulation (Figure 6.5).

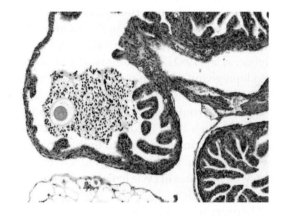

Figure 6.5   Recently ovulated ova surrounded by granulosa cells in oviduct.

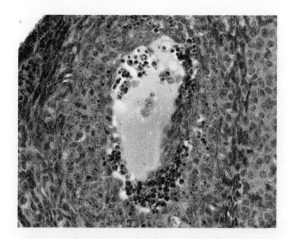

Figure 6.6 Follicle in early stages of atresia with apoptotic granulosa cells.

Most follicles do not reach the point of ovulation and either undergo attrition (primordial follicles) or atresia. Oocytes become atretic as a result of apoptosis and so in early follicular atresia fragmentation of the degenerate oocyte and granulosa cells may be seen (Figure 6.6). Remnants of atretic foillicles can be seen as shrivelled eosinophilic remnants of the zona pellucida which stain PAS positive (Figure 6.4) surrounded by theca interna cells. The remaining theca interna cells are sometimes referred to as 'interstitial glands'.

The appearance of corpora lutea varies with the stage of the oestrous cycle (Figure 6.7) and whether they are from the current or previous cycles (Figure 6.8). When first forming following ovulation the theca cells (Young and McNeilly 2010), which make up the corpora lutea, have basophilic cytoplasm and are a plump spindle shape, mitoses may be present. The centre of a newly formed corpus luteum may have an irregular fluid filled or haemorrhagic cavity. During metoestrus and dioestrus the cells of the corpus luteum become plumper and may start to accumulate fine lipid vacuoles (Greenwald and Rothchild 1968) reaching maximum size and vacuolation at dioestrus. Fibrous tissue may also be apparent in the centre of the corpus luteum at dioestrus filling what was the haemorrhagic cavity

(Westwood 2008). During proestrus the corpora lutea start to degenerate (with vacuolation of cytoplasm), nuclear fragments and neutrophils may be present. Corpora lutea from up to three previous cycles (Greenwald) may be present and are identifiable by being more eosinophilic than the corpora lutea of the current cycle and becoming paler and smaller with time.

The rete ovarii are tubular remnants of the mesonephric ducts, which can be found within the mesovarial fat or within the ovarian stroma at the hilus. The ducts may be single or multiple and lined by cuboidal to columnar epithelium, which may be ciliated. In ageing mice it is common for the ducts to become dilated or cystic (Long 2002).

6.2.3 Oviduct

The oviduct consists of four anatomical regions; intramuscular (uterotubular), isthmus, ampulla and infundibulum (Stewart and Behringer 2012) lined by a folded mucosal surface (Figure 6.9). The light microscopic appearance of the oviductal mucosa does not vary significantly with the stage of the oestrous cycle. The intramuscular portion passes through the tip of the uterine horn and is lined by low columnar cells surrounded by a connective tissue lamina propria with scattered elastic fibres and smooth muscle cells. The remaining oviduct is lined by a single epithelial layer of pseudostratified cells with centrally located nuclei. The cells are a mixture of ciliated and secretory cells with ciliated cells being more common in the infundibulum and ampulla and secretory cells in the isthmus (Yamanouchi et al. 2010; Stewart and Behringer 2012). The ampulla and isthmus have an outer wall of smooth muscle. The isthmus has a thick muscular wall and connects the intramuscular section to the ampulla and is highly coiled in the mouse, so it is common to see multiple cross-sections in standard preparations. The ampulla has an increased lumen diameter compared to the isthmus and connects to the infundibulum which ends in the fimbria in the ovarian bursa.

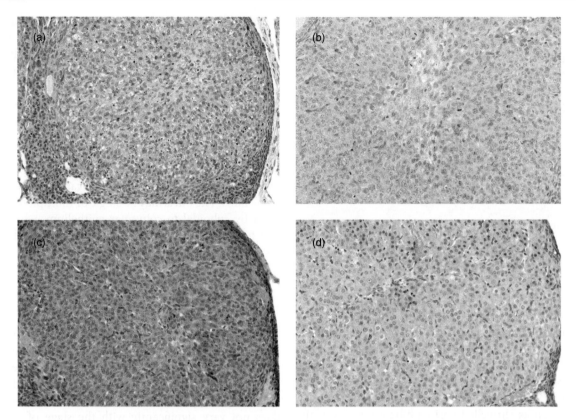

Figure 6.7   Corpora lutea at different stages of the oestrous cycle. In (a) prooestrus degenerate cells are present, (b) oestrus there may be central haemorrhage and mitsoses, during (c) metoestrus and (d) dioestrus cells increase in size and become paler and vacuolated.

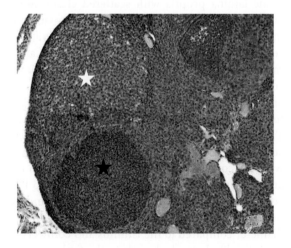

Figure 6.8   Corpora lutea from current cycle are more basophilic (black star) than those from previous cycles (white star).

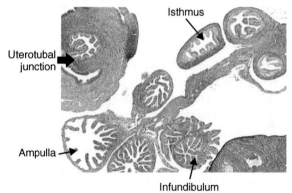

Figure 6.9   Regions of oviduct.

6.2.4   Uterus and cervix

The uterus consists of a body and two elongated horns (Chapter 1, Figure 1.36). The cranial portion of the uterine body (or corpus) is divided by a midline septum; caudally the body becomes one chamber, which is continuous with the cervix and the vagina. The uterus is composed of an outer loose connective tissue serosal layer (perimetrium), two muscular layers (myometrium) and an inner mucosal layer (endometrium). The myometrium consists of an inner circular and an outer longitudinal layer of smooth muscle fibres separated by a thin vascular connective tissue layer. In the horns and divided portion of the uterine body, the endometrium and endometrial glands are lined by a single layer of columnar epithelial cells, the appearance of which varies with the stage of the cycle

(Figure 6.10). The epithelial lining of the undivided portion of the uterine body and cervix is stratified squamous and continuous with the vaginal lining. At the start of oestrus the lumen of the uterus may be mildly distended with fluid, which gradually reduces with time (this normal appearance needs to be differentiated from fluid enlargement due to distal obstruction of the reproductive tract leading to gross distension due to mucometra or hydrometra (Chapter 1, Figure 1.38)). The glandular epithelium starts to degenerate which can be seen by vacuolation and apoptosis of the epithelial cells and there is an absence of mitoses; some infiltrating inflammatory cells may be present. In metoetrus the lumen becomes reduced and the lining epithelium continues to show signs of degeneration, although some mitoses may be present. In dioestrus the uterus is at its thinnest with a narrow slit-like lumen

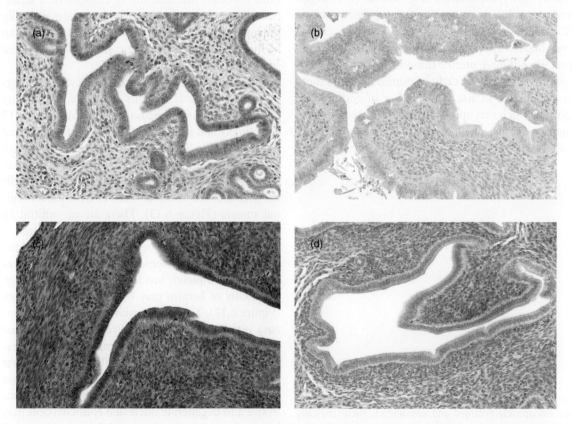

Figure 6.10   Uterine endometrium changes morphology with stage of oestrous cycle. (a) During proestrus the epithelium may contain mitoses and the stroma may be oedematous, (b) epithelium starts to vacuolated and degenerate in oestrus, (c) mitoses are seen towards the end of metoestrus, (d) the epithelium is at its lowest in dioestrus.

surrounded by a thin endometrium with condensed stromal tissue and low columnar lining cells with no evidence of degeneration although occasional mitoses may be present. In proestrous the uterine lumen becomes dilated by fluid and the lining epithelial cells become tall, columnar with frequent mitoses and the stroma may have an oedematous appearance.

A number of changes are commonly seen in the uterus of ageing mice; these include adenomyosis, endometrial hyperplasia and haemangiectasia (Creasey 2011). Adenomyosis is the presence of normal well differentiated endometrial glands and associated stroma within the myometrium, occasionally extending to the serosa, and should not be confused with neoplasia. Endometrial hyperplasia is extremely common in ageing mice, sometimes in association with ovarian cysts. At necropsy, endometrial hyperplasia is recognized as an enlarged and convoluted or segmented appearing uterus (Figure 6.11). The histological appearance can be very bizarre and easily confused with a neoplastic process, although uterine tumours are relatively uncommon in mice. In early or minimal lesions there is an increase in size and number of endometrial glands, which may appear widely separated by oedematous stroma, which in time become more densely collagenous and accompanied by a population of

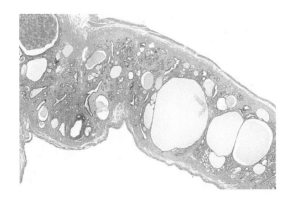

Figure 6.11 Ageing female mice commonly have varying degrees of endometrial hyperplasia. The uterus can become macroscopically enlarged and convoluted. Microscopically the appearance of hyperplasia, cysts often combined with haemangiectasis and thrombi can be mistaken for neoplasia.

neutrophils. In more advanced lesions the glands become tortuous and cystic . The epithelium may be columnar, with more basophilic cytoplasm and nuclear crowding or may be flattened in cystic areas. Dilatation of blood vessels (haemangiectasia) with associated thrombi may be seen alone or in conjunction with endometrial hyperplasia and can result in sufficient chronic blood loss to cause anaemia.

6.2.5 *Vagina*

The vagina is a tubular structure, which is continuous with the cervical canal and ends caudally at the external genital opening (vulva). The vagina has an outer layer of loose connective tissue (adventitia), two layers of smooth muscle and an inner mucosal layer. The muscular layers are arranged as an inner circular and an outer longitudinal layer. The vaginal (and cervical) mucosal lining consists of a stratified squamous epithelium which changes radically during the oestrous cycle. The cyclic changes can be monitored *in vivo* by the examination of vaginal smears (Caligoni 2009).

Oestrus is the easiest stage of the cycle to recognize in histological sections (Figure 6.12). During oestrus, the vaginal mucosal is at its thickest (10-12 cells thick) and is covered by a layer of keratinized (cornified) cells. These keratinized cells are flattened, highly eosinophilic and usually anuclear (they may be nucleated early in oestrus). These keratinized cells are shed into the lumen and can be detected on vaginal smears (Figure 6.13). There are no infiltrating neutrophils in the lamina propria or epithelium.

Metoestrus tends to be the shortest stage and the least distinct morphologically. The keratinized layer is usually completely lost from the surface although fragments of keratin may be present in the lumen (Figure 6.12). The most striking feature of the mucosa is the infiltration of neutrophils in the lamina propria and into the epithelium. This normal feature can be confused with an inflammatory process if the investigator is unaware of the cyclic changes in this tissue. The epithelium also starts to reduce in size and degenerate nucleated epithelial cells may be seen. On smears, neutrophils and degenerate epithelial cells are present.

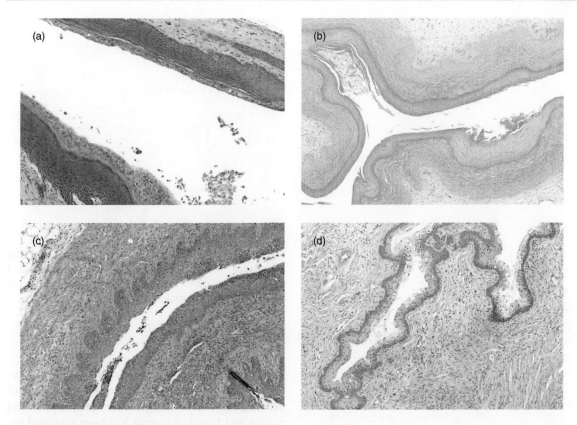

Figure 6.12   Vaginal epithelium changes morphology with stage of oestrous cycle. (a) Proestrus, (b) oestrus, (c) metoestrus, (d) dioestrus.

At the start of dioestrus, the epithelium is at its lowest height and tends to be more basophilic. Infiltrating neutrophils gradually reduce during this stage and become confined to the lamina propria. During dioestrus the epithelium starts to proliferate and mitotic figures may be present in the basal epithelial layer. At the end of dioestrus the surface layer of epithelium starts to become mucified. The cells in the mucified layer become larger and more basophilic and may contain mucin vacuoles. The mucified layer also gets thicker (2–4 cells thick) as the cycle progresses into proestrus). Vaginal smears will show large numbers of neutrophils, which have migrated through the epithelium.

During proestrus, a stratum corneum of flattened nucleated cells develops between the basal cell layers (stratum germinativum) and the mucified layer (stratum mucification). The stratum corneum gradually undergoes keratinization (cornification) leading to the appearance of a two- to three-cell thick layer of flattened highly eosinophilic cells separating the mucified layer from the basal layers. The mucified cell layer starts to be shed towards the end of proestrus so that by the start of oestrus the stratum corneum is largely exposed. Vaginal smears will show a mix of nucleated epithelial cells and some anuclear keratinized cells in proestrus.

6.2.6   *Clitoris and glands*

The clitoris is found as a small elevation on the ventral floor of the vaginal opening, projecting into the clitoral fossa. It is lined by vaginal epithelium. A small core of bone analogous to the os penis and tortuous vascular channels of erectile tissue are present dorsally. The urethra and clitoral ducts

(a)

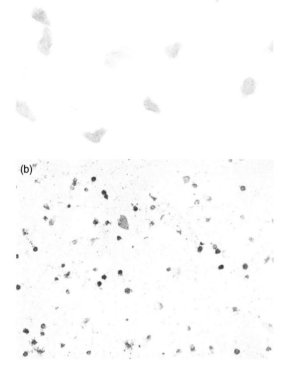

(b)

Figure 6.13   Methylene blue stained vaginal smears with (a) squames and (b) mix of neutrophils and squames.

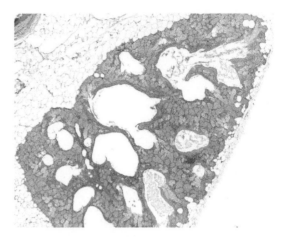

Figure 6.14   Preputial and clitoral glands have similar appearance. The photomicrograph shows a preputial gland with dilated stratified squamous epithelial lined ducts and a mononuclear inflammatory infiltrate.

6.2.7   Placenta

The placenta is never examined in standard phenotypic screens and is rarely examined in detail histologically. The comparative anatomy of the mouse and human placenta has been reviewed recently (Rendi *et al.* 2012) and only a limited overview of the fully developed placenta is presented here. The mouse has a hemochorial type of placenta where maternal blood is in contact with foetal chorion via a labyrinthine layer (humans have a similar type of placentation). The placenta is composed of maternal and embryonic-derived layers (Figure 6.15).

The maternally derived layer is the decidua, which develops from the endometrium. The decidua has two parts – the decidua basalis and decidua capsularis. The basalis has a rich vascular network and is composed of loosely adherent irregular large vacuolated cells with variably sized and shaped nuclei. The decidual cells tend to be arranged in a more compact layer closest to the junctional zone and more loosely arranged towards the maternal side. The decidua capsularis is a thin layer of flattened decidual cells that encloses the entire embryo.

empty into the clitoral fossa (Yang *et al.* 2010; Weiss *et al.* 2012).

The clitoral glands are modified sebaceous glands found ventrally just cranial to the vulva. They are rarely examined for their own sake but may be seen in sections taken around the perineal region of the mouse so it is helpful to be familiar with their histological appearance. The glands consist of ducts lined by stratified squamous epithelium and acini composed of large cuboidal eosinophilic epithelial cells and are similar to preputial glands in male mice (Figure 6.14). Cystic distension of acini and distension (ectasia) of ducts are common in ageing mice and should not be confused with neoplastic processes. The cystic conditions are usually diffuse and bilateral whereas neoplastic conditions are more commonly focal and unilateral.

The decidua is separated from the embryonic labyrinthine zone by the junctional zone. The junctional zone contains maternal blood vessels

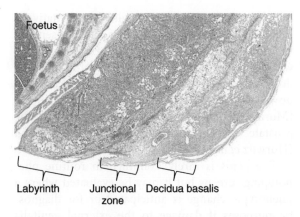

Figure 6.15 Low-power image of placenta and foetus showing major layers.

(venous drainage), trophoblast giant cells and spongiotrophoblast cells. Trophoblast cells are derived from the embryo and are generally very large with bizarrely shaped or multiple nuclei. Giant cell trophoblasts are, as their name suggests, very large – up to 100 μm in diameter and found closest to the decidua. Spongiotrophoblasts are phagocytic and may contain phagocytosed erythrocytes and are closer to the labrynthine zone (Figure 6.16).

The labrynthine zone is made up of closely opposed foetal and maternal blood channels. The foetal blood channels are lined by thin endothelial cells and contain large immature nucleated red blood cells

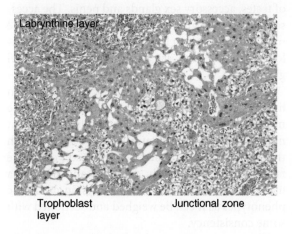

Figure 6.16 High-power image of foetal derived components of the placenta.

whereas the maternal blood channels are lined by large labrynthine trophoblast cells. Mineralization is common in this layer.

6.3   Male reproductive tract

6.3.1 *Sampling techniques*

Minimal analysis of the male reproductive system requires examination of the testes. Formalin fixation can be adequate for routine examination but as formalin and coagulant fixatives like ethanol result in significant shrinkage of the seminiferous tubules (Figure 6.17) , modified Davidson's or Bouin's fixative are often recommended for optimal testes fixation (Latendresse *et al.* 2002). Ideally, testes should be examined with regard to the spermatogenic cycle and transverse sections through the middle of the testes produce the most consistent cross sections of seminiferous tubules for this type of analysis (Moraweitz *et al.* 2004). Longitudinal sections may be necessary if lesions are expected in the polar regions. Haematoxylin and eosin stain is satisfactory for morphological examination of all male reproductive tract components but PAS stained sections are useful for assessment of acrosomal development in the testes. For basic analysis the epididymis may be sectioned with the testes, giving a transverse section through the epididymal body, but for more detailed analysis the epididymides should be carefully dissected from the testis and examined separately after fixation in formalin. Longitudinal sections allow for analysis of all regions of the epididymis. Microdissection techniques may be necessary if evaluation of the efferent ducts is required (Hess *et al.* 2000).

More extensive analysis of the male reproductive tract will require preparation of the accessory glands including the prostate, seminal vesicles and preputial glands. Seminal vesicles and coagulating gland can be examined together by taking a transverse section at the widest part (Moraweitz *et al.* 2004). Alternatively, the seminal vesicles can be sectioned longitudinally by embedding both glands dorsal surface down (Suwa *et al.* 2002).

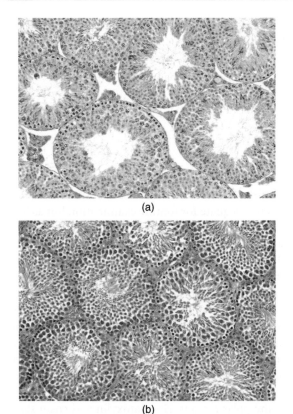

(a)

(b)

Figure 6.17   The effect of fixation on testes morphology. (a) Bouins fixative, resulting in reduced shrinkage of the germinal epithelium but increased separation of the tubules in the centre of the testes. (b) Formalin fixation can result in shrinkage of germ cells and Sertoli cells, leading to artefactual separation of the cells in the germinal epithelium even if the overall diameter of the tubules is preserved.

The mouse prostate is a complex structure and complete analysis requires examination of the dorsal, lateral and ventral lobes. The prostate and urinary bladder should be fixed together after trimming away excess fat. Formalin or paraformaldehyde are adequate for fixation. Fixing the tissue flat in a tissue cassette may make identification and dissection of the tissues for embedding easier (Suwa *et al*. 2002). Following removal of the bladder the prostate and urethra can be sectioned transversely at the distal end of the seminal vesicle and both sections embedded (Suwa *et al*. 2002; Shappell *et al*. 2004). These sections will also include the vas deferens and ampullary gland. Alternatively, following removal of the bladder and the seminal vesicles, the prostate can be flattened and embedded with the ventral surface of the ventral prostate down and the dorsal surface of the dorsal prostate down so that a horizontal section can be made (Moraweitz *et al*. 2004). Individual lobes of the prostate can be dissected separately if required (Hurwitz *et al*. 2001).

The penis is rarely examined in routine phenotyping screens but it may be evaluated where a phenotypic change is anticipated or for diagnostic purposes if damage to the external genitalia is observed. The penis can be dissected within or free from the prepuce and sectioned longitudinally, usually in a vertical plane (Rodriguez *et al*. 2011).

The bulbourethral glands are paired glands found at the base of the penis in the mouse covered by skeletal muscle. They are usually not examined in routine studies but can be excised separately or fixed whole with the penis and sectioned in cross section (Kiupel *et al*. 2000).

The preputial glands are found lateral to the penis in the mouse and can be embedded dorsal surface down and longitudinal horizontal sections prepared (Ruehl-Fehlert *et al*. 2004).

6.3.2   *Anatomy and histology*

The reproductive system of the male mouse consists of testes, accessory sex glands and penis. The accessory sex glands are particularly complicated in the mouse and consist of large seminal vesicles, a multi-lobed prostate, ampullary glands and bulbourethral glands.

The paired testes are found in the scrotum in life but due to the open inguinal canals of the mouse are commonly found intra-abdominally at necropsy (Chapter 1, Figure 1.40). Associated with the testes and epididymides is a white adipose tissue fat pad, which is readily dissected and may be a useful tissue to evaluate when analysing metabolic phenotypes as it can be weighed and sectioned with some consistency.

The testes are encased in a fibrous capsule, the tunica albuginea, and consist of tightly packed seminiferous tubules and interstitial (Leydig) cells (Figure 6.19). The seminiferous tubules are highly convoluted and so multiple cross sections will be seen for each tubule in histological sections. Each tubule is surrounded by a layer of myoid cells and a basement membrane. The seminiferous tubules end in the ducts of the rete testes, which is located below the capsule. As the seminiferous tubules approach the rete they may be lined only by Sertoli cells and rete tubules themselves are lined by a low cuboidal epithelium (Figure 6.18). These anatomical features should not be confused with seminiferous tubular atrophy, which can occur sporadically as a background finding in mice of any age (Figure 6.21). Blood vessels are normally present in the capsule adjacent to the rete and so can help confirm the position of the rete (Figure 6.18). Rete dilation and hyperplasia of the rete epithelium are common background findings in older mice.

The testes are prone to artefact due to fixation and handling. Formalin fixation can result in shrinkage of germ cells and Sertoli cells leading to artefactual separation of the cells in the germinal epithelium even if the overall diameter of the tubules is preserved (Chapin *et al.* 2004) (Figure 6.14). Alternative fixatives such as Bouins and modified Davidsons result in reduced shrinkage of the germinal epithelium and are therefore considered preferable for evaluation of spermatogenesis but may cause increase

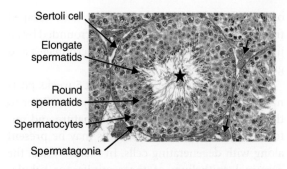

Figure 6.19 Cross-section of seminiferous tubule in early stage of spermatogenic cycle (star). Interstitial cells are present between tubules (short arrows).

separation of the tubules from each other particularly in the centre of the testes unless resin sections are prepared (Latendresse *et al.* 2002; Chapin *et al.* 2004). Compression of the testes during dissection can lead to dissociation of germinal epithelium and the appearance of clumps of displaced cells in the tubular lumen which should not be confused with pathological shedding of the epithelium (Figure 6.20).

A range of background and incidental findings are found in young and ageing mice (Creasey 2011). Mineralization of tubules with or without clumps of sperm (sperm stasis) are common in ageing mice. Sperm granulomata may also be seen in the testes

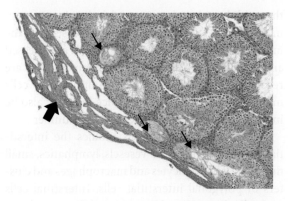

Figure 6.18 Sertoli-lined tubules (thin arrow) are normal close to the rete. Blood vessels are commonly found in the capsule associated with the rete (thick arrow).

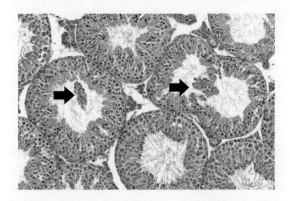

Figure 6.20 Compression of the testes during dissection can lead to dissociation of germinal epithelium and clumps of displaced cells in the tubular lumen (arrows), which should not be confused with pathological shedding of the epithelium.

or epididymides, where sperm that have escaped from the seminiferous tubules are surrounded by an inflammatory reaction involving macrophages and giant cells.

Male mice reach puberty at about 5 weeks post-natally (sperm present in the epididymis). Prior to this, the germinal epithelium is immature and mitoses of germinal and Sertoli cells may be present along with degenerating cells. In mature mice, the germinal epithelium of the seminiferous tubules consists of Sertoli cells and developing sperm cells. Sertoli cells line the basement membrane of the tubules and are joined by tight junctions which play an important role in the physical blood testes barrier (Mital *et al.* 2011). The Sertoli cells look roughly triangular in standard sections but their cytoplasm actually extends to the seminiferous tubular lumen surrounding the germ-cell populations (Russell *et al.* 1990a). Sperm cells mature as they progress towards the lumen with the most immature cells, the spermatagonia, located basally and the maturing spermatocytes, spermatids and spermatozoa being present in the layers moving towards the lumen. It takes 35 days for sperm to mature in the seminiferous tubules and 12 discrete stages are defined morphologically in mice (Russell *et al.* 1990b) by the presence of populations of spermatocytes and spermatids (one generation of spermatocytes are present in routine sections in stages I – VII (Figure 6.19) whereas two generations are usually visible in stages IX–XII (Figure 6.21). Round and elongate spermatids are present in stages I–VIII but only elongate spermatids in stages IX-XII) at different stages of differentiation and the development of the acrosome, which can be visualized in PAS-stained sections in round spermatids (Figure 6.22). The percentage of tubules in each stage can be analysed (Russell *et al.* 1990b; Creasey 1997; Ahmed and De Rooij 2009) if necessary to look for subtle alterations in spermatogenesis. However for routine screening purposes, examination of H&E-stained sections, with an appreciation that most testicular tubules are lined by a thick germinal epithelium, that there are tubules at different stages with elongated spermatids in some tubules and sperm in

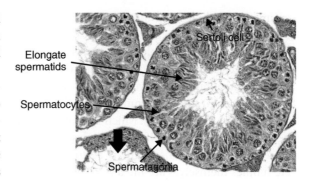

Figure 6.21   Late-stage tubule with an adjacent atrophic tubule lined by Sertoli cells (thick arrow).

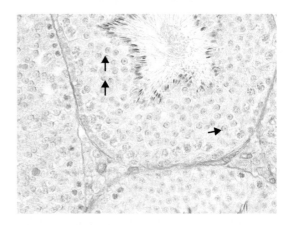

Figure 6.22   Acrosomes can be highlighted using a PAS stain (arrows).

the epididymides gives a good indication of normal function. Residual bodies can be seen at the lumen (Figure 6.23) when spermatozoa are released and further down into the epithelium as they are resorbed by the Sertoli cells. Multinucleate cells may be present in immature mice but may also be indicative of pathology if increased in number.

Between the seminiferous tubules the interstitial tissue contains blood vesesels, lymphatics, small numbers of lymphocytes and macrophages and clusters of polygonal interstitial cells. Interstitial cells generally have abundant eosinophilic cytoplasm, which may be variably vacuolated, and large central nuclei (Figure 6.19).

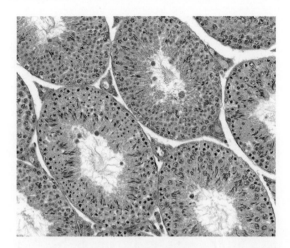

Figure 6.23 Large residual bodies are common, particularly in young mice and need to be differentiated from shed germinal epithelium.

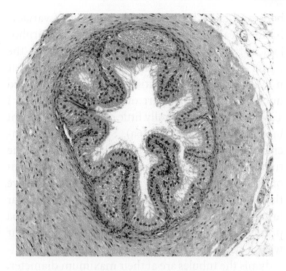

Figure 6.25 Vas deferens lined by ciliated epithelium.

The pathway of travel for sperm is from the seminiferous tubules through the rete to efferent ductules (Figure 6.24), which empty into the head of the epididymis and then exiti the tail of the epididymis via the vas deferens (Figure 6.25).

Efferent ductules are occasionally seen in cross section either embedded in fat close to the testes or as multiple cross sections of the coiled region in a fibrous capsule continuous with that of the epididymis. The ducts are lined by a single layer of variable cuboidal to columnar epithelium and are surrounded by myoid cells. A proportion of the cells are ciliated.

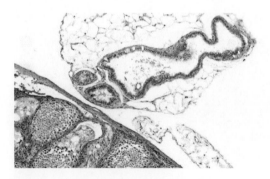

Figure 6.24 Efferent ducts can sometimes been seen adjacent to the rete of the testes.

The epididymides are closely adherent to the testes along their long axis but, with care, can be dissected free. The epididymis is surrounded by a fibrous capsule enclosing the tubules, which are surrounded individually by interstitial tissue containing blood vessels, lymphatics, connective tissue and small numbers of scattered macrophages and lymphocytes. The blood vessel density is highest in the most proximal portion on the head region (Takano 2007). Small foci of lymphocytes are common and sperm granulomata can also be seen incidentally. The epididymal tubules are lined by a single layer of epithelial cells aligned on a basement membrane, which is surrounded by concentric layers of myoid cells that is more pronounced in the tail than the head (Mital *et al.* 2011). In mature animals, spermatozoa are present in the lumen and occasional cell debris is considered normal. An absence of spermatozoa suggests immaturity or a pathological process.

The epithelial cell morphology and function varies along the length of the epididymis and while, macroscopically, the epididymis is divided into three sections, five morphological segments can be distinguished histologically (Takano 2007) (Figure 6.26). In general, the tubular diameter and lumen is smallest in the head (caput) and widens through the body and tail. The epithelial lining is most variable in the head region with three distinct morphological types being recognizable. In the head, the lining cells are ciliated, tall columnar and appear multilayered (pseudostratified) with basally located nuclei. In the initial part of the head (Segment I Takano) there is generally little or no sperm present in the lumen (Figure 6.26a). Sperm numbers increase through the second and third segments of the tubule in the head region. In the second segment of the head the cytoplasm may be vacuolated above the nuclei. In the body of the epididymis the tubular lumen is wider, there are increased spermatozoa present and the epithelial cells are lower columnar with a dense brush border. In the tail of the epididymis the tubules are at their maximum diameter, densely packed with spermatozoa and lined by low cuboidal epithelium with round to flattened nuclei (Figure 6.26b).

The vas deferens is not routinely examined in standard phenotyping or toxicity screens but may be seen occasionally with sections of epididymis or prostate. The tubule is lined by tall pale staining columnar cells with central nuclei. Long cilia project into the lumen and are mixed with spermatozoa in mature animals. The vas has a thick muscular wall, which consists of three smooth muscle layers, thin inner and thick outer longitudinal layers and a central circular layer. The loose connective tissue adventitia surrounding the muscle layers contains blood vessels (Figure 6.25).

The ampullary glands are expansions of the terminal vas deferens and empty via single ducts into the vas deferens proximal to the seminal vesicles. The acini are surrounded by a relatively thick layer of fibrovascular stroma, which helps to distinguish the gland from the prostatic lobes. The epithelial lining is folded and lined by cuboidal epithelium with basal nuclei. The luminal secretion is densely eosinophilic, which separates to give characteristic irregular holes (Figure 6.27).

The bilateral seminal vesicles are sacculated glands (Figure 6.28) lying dorsolateral to the urinary bladder (Chapter 1, Figure 1.41), which empty via ducts into the membranous urethra at the seminal collicle (Radovsky *et al.* 1999).

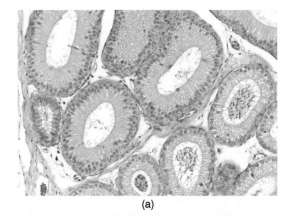

(a)

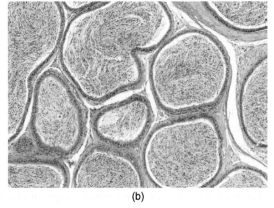

(b)

Figure 6.26 The epithelium height, lumen size and presence of spermatozoa varies between the head (a) and tail (b) of the epididymis.

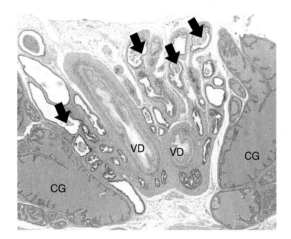

Figure 6.27 Ampullary glands (arrows) empty into the distal vas deferens (VD). Coagulating glands (CG).

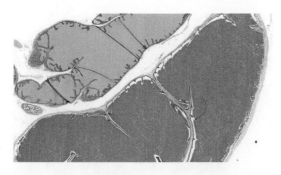

Figure 6.28 Seminal vesicle (below) contain brightly eosinophilic secretion compared to the paler secretion of the coagulating gland (above).

The glands can become very distended with secretion, which may become hardened and inspissated in older animals. The coagulating gland, which forms the anterior part of the prostate, lies adjacent to the medial aspect of the seminal vesicles (Figure 6.28). The seminal vesicles are surrounded by a smooth muscle layer and lined by a single layer of columnar epithelium, which is intricately folded over a loose connective tissue lamina propria. The branching folds of the epithelium may give the impression that the epithelium is hyperplastic when the glands are contracted but are a normal feature and so should not be mistaken for a proliferative change. The epithelial cells have round to oval basally placed nuclei and numerous bright eosinophilic granules in the apical cytoplasm. The luminal secretion of the seminal vesicles is also brightly eosinophilic, which helps to distinguish it from the coagulating gland.

The prostate of the mouse consists of anterior, lateral, dorsal and ventral lobes, each of which are surrounded by an individual mesothelial lined capsule. The acini within the lobes are separated by sparse, loose connective tissue and surrounded by thin fibromuscular layers composed of smooth muscle fibres and collagen. The anterior prostate or coagulating gland is lined by a single layer of epithelial cells, which follow a folded or papillary pattern. The epithelial cells are low columnar with centrally placed nuclei and bright eosinophilic granules in the apical cytoplasm. The luminal secretion is pale

eosinophilic in contrast to that of the adjacent seminal vesicles. The acini of the dorsal prostate are small in diameter compared to the other lobes and lined by columnar epithelium, which is arranged in small folds with centrally placed nuclei and pale eosinophilic luminal secretion (Figure 6.29a). The tubules of the lateral prostate have a larger diameter lumen filled with granular pale eosinophlic secretion and lined by cuboidal epithelium, which has only occasional folds, with basal nuclei (Figure 6.29a). The ventral prostate has variable-sized acini with a less folded epithelium than the dorsal prostate, comprising cuboidal to columnar epithelium with small

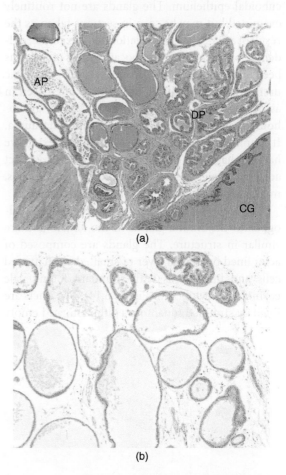

(a)

(b)

Figure 6.29 The mouse prostate has four lobes – (a) the anterior prostate (also known as the coagulating gland CG), dorsal prostate (DP), lateral prostate (LP) and (b) the ventral prostate.

basal nuclei and pale granular luminal secretion (Figure 6.29b).

The paired bulbourethral glands are found adjacent to the urethra at the base of the penis underneath the bulbocavernosus muscle (skeletal muscle). The ducts of the glands empty into the urethra proximal to the urethral diverticula. The urethral diverticula is not present in all mouse strains. The bulbourethral glands have a tail and body section and are surrounded by smooth muscle. The glands are made up of acini of cuboidal epithelium with pale foamy cytoplasm and small basally located nuclei (Figure 6.30). The ducts are lined by cuboidal epithelium. The glands are not routinely examined but may be seen in sections taken as the result of a diagnostic observation – for example, reddening of the overlying bulbocavernosus muscle has been noticed at necropsy secondary to haemorrhagic urethritis. Haemorrhagic urethritis is often seen in association with balanoposthtitis in young, group-housed mice, as a result of fighting-induced trauma. It manifests as acute inflammation with extensive haemorrhage at the penile flexure which may affect adjacent structures including corpus cavernosus, muscle and bulbourethral glands (Figure 6.31).

The preputial glands in the male mouse are analogous to the clitoral glands in the female and are similar in structure. The glands are composed of acini lined by a basal layer of small, dark, flattened cells and lining cells with abundant foamy pale eosinophilic cytoplasm (Figure 6.14). The ducts are lined by stratified squamous epithelium and empty

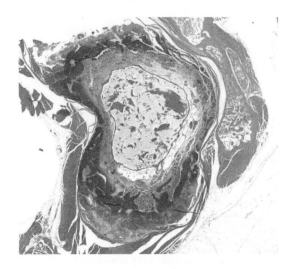

Figure 6.31 Haemorrhagic urethritis with extensive haemorrhage at the penile flexure affecting adjacent structures including corpus cavernosus, muscle and bulbourethral glands.

into the preputial space near the external opening. Preputial gland inflammation and abscessation is common in some mouse colonies and may be noticed as a swelling around the perineal region (Chapter 1, Figure 1.16). Cystic dilatation of the ducts and acini is also common in older animals.

The mouse penis consists of a body and glans, which is covered by the prepuce. The detailed anatomy and histology of the mouse penis has recently been described (Yang *et al.* 2010; Rodriguez *et al.* 2011) but as this structure is not routinely examined a brief overview is given below. The body of the penis starts distal to the prostate and ends at a right-angled bend where the glans starts. The urethra is lined by transitional epithelium, passes from the bladder to the tip of the penis and is divided into a thin-walled membranous proximal portion, as it passes through the prostate, and a distal penile segment. The urethra is surrounded by cavernous, thin-walled blood vessels (erectile tissues) and an os penis is present in the glans dorsal to the urethra (Rodriguez *et al.* 2011). The os penis is made of bone proximally and ends distally in the tip of the glans as a fibrocartilagenous structure – the male urogenital mating protuberance (MUMP) (Figure 6.32). The

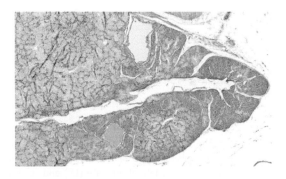

Figure 6.30 Bulbourethral glands.

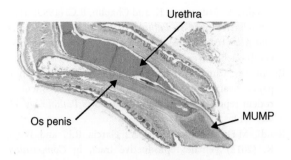

Figure 6.32 Longitudinal section of penis. An eosinophilic colloidal plug is present in the urethra. This is a common agonal finding and can be distinguished from an obstruction by the absence of associated inflammation in the urethral epithelium.

glans is covered in a keratinized, stratified squamous epithelium, which has spinous processes proximally, a feature not present in the epithelium covering the MUMP region. The prepuce is lined by stratified squamous epithelium.

The presence of an eosinophilic colloidal plug in the penile urethra is common and is thought to be an agonal release of secretion as a result of euthanasia. This incidental agonal change needs to be differentiated from an obstructive urethral plug, which commonly occurs secondary to urethritis or balanoposthitis and may result in the death of the

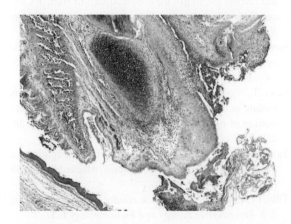

Figure 6.33 Balanoposthitis can result from trauma to the external genitalia, particularly in group-housed mice and can lead to blockage of the urinary tract or ascending infections.

animal. Obstructive plugs are usually paler in colour than the agonal plugs and are usually accompanied by infiltrating inflammatory cells within the plug plus inflammation of associated structures.

Balanoposthitis or inflammation of the penis and surrounding tissue is common in young group-housed mice of some strains (Figure 6.33). The inflammation may extend up the urinary tract, leading to cystitis, pyelonephritis and inflammation of the various accessory glands.

References

Ahmed, E.S. and de Rooij, D.G. (2009) Staging of mouse seminiferous tubule cross-sections. *Methods in Molecular Biology* **558**, 263–277.

Blackburn, M.R. (2000). Examination of normal and abnormal placentation in the mouse. *Methods in Molecular Biology* **136**, 185–193.

Byers, S.L., Wiles, M.V., Dunn, S.L., Taft, R.A. (2012) Mouse estrous cycle identification tool and images. *PLoS ONE* **7** (4), e35538. doi:10.1371/ journal.pone.0035538.

Caligoni, C. (2009) Assessing reproductive status/stages in mice. *Current Protocols in Neuroscience* **48**, A.41.1–A.41.11.

Champlin, K.A., Dorr, D.L. and Gates, D.A. (1973) Determining stage of the estrous cycle in the mouse by appearance of the vagina. *Biology of Reproduction* **8**, 491–494.

Chapin, R.E., Ross, M.D. and Lamb, J.C. (1984) Immersion fixation methods for glycol methacrylate-embedded testes. *Toxicologic Pathology* **12**, 221–227.

Creasey, D. (1997) Evaluation of testicular toxicity in safety evaluation studies: the appropriate use of spermatogenic staging. *Toxicologic Pathology* **25**, 119–131.

Creasey, D. (2011) Reproduction of the rat, mouse, dog, non-human primate and minipig, in *Background Lesion in Laboratory Animals. A Color Atlas* (ed. E.F. Mcinnes), Saunders, Edinburgh.

El-Mestrah, M., Castle, P.E., Borosa, A. and Kan, F.W.K. (2002) Subcellular distribution of ZP1, ZP2, and ZP3 glycoproteins during folliculogenesis and demonstration of their topographical disposition within the zona matrix of mouse ovarian oocytes. *Biology of Reproduction* **66**, 866–876.

Greenwald, G.S. and Rothchild, I. (1968) Formation and maintenance of corpora lutea in laboratory animals. *Journal of Animal Science* **27**, 139–161.

Hess, R.A., Bunick, D., Lubahn, D.B. *et al.* (2000) Morphologic changes in efferent ductules and epididymis in estrogen receptor – a knockout mice. *Journal of Andrology* **21**, 107–121.

Hurwitz, A.A., Foster, B.A., Allison, J.P. *et al.* (2001) The TRAMP mouse as a model for prostate cancer. *Current Protocols in Immunology* **45**, 20.5.1–20.5.23

Kent, H.A. (1960) Polyovular follicles and multinucleate ova in the ovaries of young mice. *Anatomical Record* **137**, 521–524.

Kittel, B., Ruehl-Fehlert, C., Moraweitz, G. *et al.* (2004) Revised guides for organ sampling and trimming in rats and mice – Part 2. *Experimental Toxicologic Pathology* **55**, 413–431.

Kiupel, M., Brown, K.S. and Sundberg, J.P. (2000). Bulbourethral (Cowper's) gland abnormalities in inbred laboratory mice. *Journal of Experimental Animal Science* **40**, 178–188.

Latendresse, J.R., Warbrittion, A.R., Jonassen, H., Creasy, D.M. (2002) Fixation of testes and eyes using a modified Davidson's fluid: comparison with Bouin's fluid and conventional Davidson's fluid. *Toxicologic Pathology* **30**, 524–533.

Li, S. and Davis, B. (2007) Evaluating rodent vaginal and uterine histology in toxicity studies. *Birth Defects Research Part B: Developmental and Reproductive Toxicology* **80**, 246–252.

Long, G.G. (2002) Apparent mesonephric duct (rete anlage) origin for cysts and proliferative epithelial lesions in the mouse ovary. *Toxicologic Pathology* **30**, 592–598.

Mital , P., Hinton, B.T. and Dufour, A.M. (2011) The blood-testis and blood-epididymis barriers are more than just their tight junctions. *Biology of Reproduction* **84**, 851–858.

Morawietz ,G., Ruel-Fehlert, C., Kittel, B. *et al.* (2004) Revised guides for organ sampling and trimming in rats and mice – Part 3. *Experimental Toxicologic Pathology* **55**, 433–449.

Myers, M., Britt, K.L., Wreford, N.G.M. *et al.* (2004) Methods for quantifying follicles within the mouse ovary. *Reproduction Research* **127**, 569–580.

Nelson, J.F., Felicio, L.S., Randall, P.K. *et al.* (1982) A longitudinal study of estrous cyclicity in aging C57Bl/6J mice: 1. Cycle frequency, length and vaginal cytology. *Biology of Reproduction* **27**, 327–339.

Nelson, J.F., Karelus, K., Felicio, L.S. and Johnson, T.E. (1990) Genetic influences on the timing of puberty in mice. *Biology of Reproduction* **42**, 649–655.

Ono, M. and Harley, V.R. (2013) Disorders of sex development: new genes, new concepts. *Nature Reviews Endocrinology* **9**, 79–91.

Pritchett, K.R. and Taft, R.A. (2007) Reproductive biology of the laboratory mouse, in *The Mouse in Biomedical Research: Normative Biology, Husbandry, and Models*, Volume 3 (ed. J.E. Fox), Academic Press, Burlington.

Radovsky, A., Mitsumori, K. and Chaplin, R.C. (1999) *Male reproductive tract, in Pathology of the Mouse: Reference and Atlas* (eds R.R. Maronpot, G.A. Boorman and B.W. Gaul) Cache River Press, Clearwater.

Regan, K.S., Cline, M., Creasy, D. *et al.* (2005) STP position paper: ovarian follicular counting in the assessment of rodent reproductive toxicity. *Toxicologic Patholology* **33**, 409–412.

Rendi, M.H., Muehlenbachs, A., Garcia R.L. and Boyd, K. (2012) Female reproductive tract, in *Comparative Anatomy and Histology a Mouse and Human Atlas* (eds P.M. Treuting and S.M. Dintzis), Academic Press, Amsterdam, pp. 273–284.

Rodriguez, E., Weiss, D.A., Yang. J.H. *et al.* (2011) New insights on the morphology of adult mouse penis. *Biology of Reproduction* **85**, 1216–1221.

Ruehl-Fehlert, C., Kittel, B., Moraweitz, G. *et al.* (2003) Revised guides for organ sampling and trimming in rats and mice – Part 1. *Experimental Toxicologic Pathology* **55**, 91–106.

Russell, L.D., Ettlin, R.A., Sinha Hikim, A.P. and Clegg, E.D. (1990a) The sertoli cell, in *Histological and Histopathological Evaluation of the Testis*, Cache River Press, Clearwater, pp. 29–35.

Russell, L.D., Ettlin, R.A., Sinha Hikim, A.P. and Clegg, E.D. (1990b) Staging for the mouse, *in Histological and Histopathological Evaluation of the Testis*, Cache River Press, Clearwater, pp. 119–161.

Shappell, S.B., Thomas, G.V., Roberts, R.L. *et al.* (2004) Prostate pathology of genetically engineered mice: definitions and classification. The consensus report from the Bar Harbor meeting of the Mouse Models of Human Cancer Consortium Prostate Pathology Committee. *Cancer Research* **64**, 2270–2305.

Stewart, C.A., Behringer, R.R. (2012) Mouse oviduct development. *Results and Problems in Cell Differentiation* **55**, 247–262.

Suwa, T., Nyska, A., Haseman, J.K. *et al.* (2002) Spontaneous lesions in control B6C3F1 mice and recommended sectioning of male accessory sex. *Toxicologic Pathology* **30**, 228–234.

Takano, H. (2007) Histological division of mouse epididymis based on regional differences. *Hirosaki Medical Journal* **59** (Suppl), S292–s301.

Tilly, J.L. (2003) Ovarian follicle counts – not as simple as 1, 2, 3. *Reproductive Biology and Endocrinology* **1**, 11.

Weiss, D.A., Rodriguez, E. Jr, Cunha, T. *et al.* (2012) Morphology of the external genitalia of the adult male and female mice as an endpoint of sex differentiation. *Molecular and Cellular Endocrinology* **354**, 94–102.

Westwood, F.R. (2008) The female rat reproductive cycle: a practical histological guide to staging. *Toxicologic Pathology* **36**, 375–384.

Yamanouchi, H., Umezu, T. and Tomooka, Y. (2010) Reconstruction of oviduct and demonstration of epithelial fate determination in mice. *Biology of Reproduction* **82**, 528–533.

Yang, J.H., Menshenina, J., Cunha, G.R. *et al.* (2010) Morphology of mouse external genitalia: implications for a role of estrogen in sexual dimorphism of the mouse genital tubercle. *Journal of Urology* **184**, 1604–1609.

Young, J.M. and McNeilly, A.S. (2010) Theca; the forgotten cell of the ovarian follicle. *Reproduction* **140**, 489–504.

Chapter 7
Endocrine system

Ian Taylor

Huntingdon Life Sciences, Eye, Suffolk, UK

7.1 Introduction

The endocrine system is a collection of discrete organs and cells distributed throughout the body whose role is to maintain homeostasis among different organs of the body. They act by secreting their cellular products (hormones) directly into the bloodstream where they act on cells at distant sites (endocrine effects); at nearby sites (paracrine effects); or on the secreting cell itself (autocrine effects). The function of the endocrine glands is usually controlled by negative feedback loops, where the release of the hormone is controlled by circulating levels of the factors released by the target organs, or plasma levels of the secreted hormone itself. The controlling mechanisms involve interactions between the central nervous system, the endocrine glands and their target organs, so disruption of one component in the homeostatic axis can have knock-on effects on the endocrine gland, sometimes leading to proliferative changes in cellular components of the endocrine gland. A breakdown in the normal homeostatic mechanisms of the endocrine system,

as a result of ageing, can lead to the hyperplastic or neoplastic changes commonly seen in senile rodents.

The anatomy and histology of the pituitary, adrenal, thyroids and parathyroids are presented below, but there are several sources of information (some of which are available online) that provide details of the basic anatomy of the mouse (Cook 1965, 1983; Hummel *et al.* 1966) and provide guidance on general necropsy and histology practices, which include descriptions of procedures to follow for the dissection and trimming of these tissues (Ruehl-Fehlert *et al.* 2003; Kittel *et al.* 2004; Knoblaugh *et al.* 2012). The endocrine portion of the pancreas is discussed in Chapter 3 with the exocrine pancreas.

7.2 Adrenals

7.2.1 *Background and development*

The paired adrenal glands of the mouse are composed of an outer cortex with two distinct layers,

A Practical Guide to the Histology of the Mouse, First Edition. Cheryl L. Scudamore.
© 2014 John Wiley & Sons, Ltd. Illustrations © Veterinary Path Illustrations, unless stated otherwise.
Published 2014 by John Wiley & Sons, Ltd. Companion Website: www.wiley.com/go/scudamore/mousehistology

the zona glomerulosa and the zona fasciculata, and the inner medulla composed of chromaffin cells and ganglion cells. There is no discernible zona reticularis in the mouse adrenal (Nyska and Maronpot 1999). The mineralocorticoid aldosterone is produced by cells of the zona glomerulosa and is responsible for maintenance of electrolyte and fluid homeostasis. Glucocorticoids are produced by the zona fasciculata. The major glucocorticoid hormone in mice is corticosterone, which is involved in carbohydrate, lipid and protein metabolism, immune functions and stress responses. Release of corticosterone is under the control of adrenocorticotropic hormone (ACTH) from the adenohypophysis, which is itself controlled by release of corticotropin-releasing hormone from the hypothalamus. The medulla is responsible for the production of epinephrine and norepinephrine, the release of which is controlled by the sympathetic nervous system. They act to reinforce the actions of the sympathetic nervous system and are crucial in the flight-or-fight response.

The cortical anlage appears on day 11 of gestation as an area of budding from the coelomic epithelium and migrates dorsally from a position medial to the developing gonads to a position between the mesonephros and the aorta. Sympathoblasts, originating from the neural crest ectoderm, migrate towards the cortical anlage, and they are in close proximity by day 12. By day 14 the medullary cells are enveloped in the developing adrenal gland, and cortical capillaries are formed. The gland grows from day 15, when there is a prominent capsule, to birth, at which time it is fully functional (Sass 1983; Nyska and Maronpot 1999; Bielinska *et al.* 2006).

7.2.2   Sampling techniques

The adrenals in the mouse are small, cream/pale tan in colour and can be hard to identify, so some authors recommend leaving them within the perirenal fat pad attached to the kidney during fixation and processing (Chapter 1). However, weighing of the adrenals requires them to be removed from the kidneys, and the surrounding fat trimmed away.

Care needs to be taken not to damage the adrenal capsule during this process. The small size of the adrenals also makes them difficult to cut so that adequate sections of medulla are present. In preparation for processing, mouse adrenals should be retained *in toto* in cassettes with biopsy sponges to ensure that they are not lost during processing. To improve sectioning, mouse adrenals are best embedded longitudinally *in toto* without other tissues and sectioned to reach the optimal level, which presents the largest cut surface (Kittel *et al.* 2004) (Figure 7.1). It is possible to embed the adrenals attached to the cranial pole of the kidney but this makes it hard to produce adequate sections of both tissues demonstrating all relevant areas (cortex, medulla and papilla of the kidney, and cortex and medulla of the adrenal).

7.2.3   Artefacts

Care is needed in sectioning the adrenal to ensure that the medulla is presented on the section. Tangential sections can give the appearance of variations in the proportion of cortex to medulla. Extension of

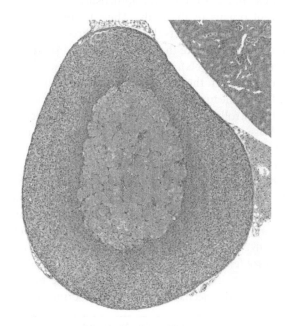

Figure 7.1   Normal cross section of mouse adrenal showing cortex and medulla.

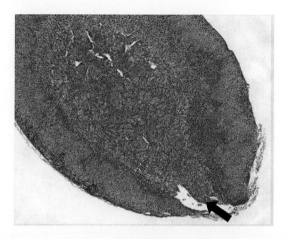

Figure 7.2 Section of adrenal illustrating extension of medullary cells along hilus towards outer cortex (arrow).

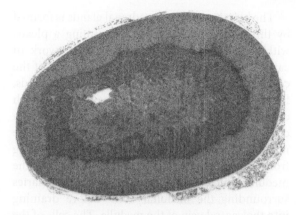

Figure 7.3 X zone of a female CD-1 mouse appearing as a distinctive layer of basophilic cells between the zona fasciculata and the medulla.

medullary cells along the hilus through the cortex towards the outer cortex is a normal feature and does not represent a proliferative change (Capen *et al.* 2001, Frith *et al.* 2007) (Figure 7.2).

7.2.4   Anatomy and histology of the adrenals

The paired adrenal glands are ovoid in shape and located adjacent to the anterior pole of the kidneys. The left gland is typically heavier than the right; they are slightly larger in females than in males and the female adrenal is paler because of the presence of more lipids. The outer zona glomerulosa is composed of small basophilic cells below the cortical capsule, and the more eosinophilic cells of the zona fasciculata form columns extending towards the medulla (Nyska and Maronpot 1999; Bielohuby *et al.* 2007). The X-zone is a unique feature of the mouse adrenal; a layer of basophilic cells at the corticomedullary region that forms postnatally and is fully formed at weaning (Figure 7.3). In males the X-zone degenerates rapidly at puberty without vacuolation; in females the zone undergoes slow vacuolar degeneration and can be present for several weeks (Figure 7.4). The thickness and persistence of the X-zone can vary with strain of mouse. The zone disappears rapidly during pregnancy (Rosol *et al.* 2001).

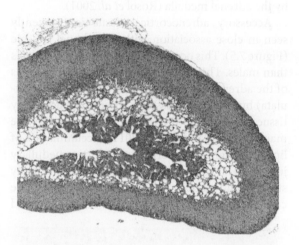

Figure 7.4 Vacuolation of the x-zone in a female cd-1 mouse.

The medulla is composed of a homogeneous population of polyhedral chromaffin cells arranged in variably sized packets, ganglion cells, venules and capillaries, with conspicuous sinusoids. The cytoplasm of the chromaffin cells is finely granular and slightly more basophilic than the cells of the cortex. Chromaffin cells stain when labeled with antibodies for tyrosine hydroxylase, chromogranin and synaptophysin. The ganglion cells are scattered randomly within the medulla (Nyska and Maronpot 1999; La Perle and Jordan 2012).

The blood supply to the adrenal glands is formed by three groups of arteries, constituting a plexus within the capsule. An anastomizing network of capillary sinusoids form the vascular system of the cortex, which descends between the cortical cords, forming a second plexus in the lower cortex, which drains into small venules in the medulla, which converge into the central vein of the medulla. The medulla is supplied by small arterioles descending from the capsular plexus through the cortex into the medulla, which ramifies into a network of capillaries surrounding the medullary cells before draining into the central vein of the medulla. The cells of the medulla are therefore exposed to fresh arterial blood as well as blood rich in corticosteroids, which has an important influence on the synthesis of hormones by the adrenal medulla (Rosol *et al.* 2001).

Accessory adrenocortical tissue is frequently seen in close association with the adrenal capsule (Figure 7.5). This occurs more commonly in females than males. They usually exhibit two distinct zones of the adrenal cortex (zona glomerulosa and fasciculata) but medullary tissue is absent. The accessory tissue can undergo the same aging changes as the main adrenal gland (Taylor 2011). Accumulation of lipogenic pigment, or ceroid, at the corticomedullary junction is a common age-related change in many strains and stocks of mice, occurring more commonly in females than males (Figure 7.6 and

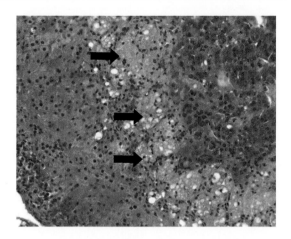

Figure 7.6 Corticomedullary pigment (arrows) in a control two year old female mouse.

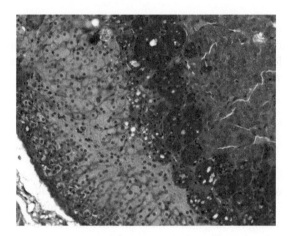

Figure 7.7 PAS positive staining of corticomedullary pigment shown in Figure 7.6.

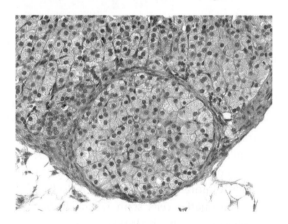

Figure 7.5 Accessory adrenocortical tissue. These tissues have a normal cortical structure, and zona glomerulosa and fasciculata can be evident. Medullary chromaffin cells are never seen.

Figure 7.7). Pigment is deposited in cortical cells and macrophages, and stains positively with the Periodic Acid Schiff (PAS) stain (Taylor 2011). The adrenal glands are described as a common site for deposition of amyloid in different strains of mice, particularly the CD-1 mouse (Brayton 2007; Frith *et al.* 2007). However, the occurrence of amyloidosis in CD-1 mice has shown time-related changes and can vary considerably between animal suppliers (Maita *et al.* 1988; Engelhardt *et al.* 1993; Taylor 2011).

Other than hyperplasia of cortical subcapsular cells, proliferative changes including neoplasms are a relatively rare occurrence in most strains of mice (Nyska and Maronpot 1999; Brayton 2007). Sub-capsular cell (spindle cell) hyperplasia is a common age-related change in the cortex in a number of strains of mice, more often in females than males. It rarely occurs before 3 months of age, but the number of subcapsular cells increases with age, forming focal, wedge-shaped accumulations, or proliferating diffusely along the outer cortex. Two types of subcapsular cell have been identified: type-A cells, fusiform in shape with scant basophilic cytoplasm, and type-B cells, larger, round cells with eosinophilic or vacuolated cytoplasm (Figures 7.8 and 7.9). There

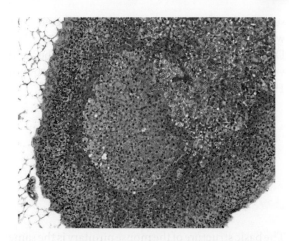

Figure 7.10 Focal cortical hypertrophy. Discrete focus of enlarged cells with eosinophilic cytoplasm.

are strain differences in the occurrence of these changes, and also in the association of mast cells with subcapsular cell hyperplasia (Kim *et al.* 2000; Taylor 2011). Sex hormones also play a part in the development of subscapular cell hyperplasia, as gonadectomy leads to an increased incidence in males (Bernichtein *et al.* 2009).

Focal hypertrophy/hyperplasia of the cells of the zona fasciculata is also seen occasionally (Nyska and Maronpot 1999; Capen *et al.* 2001) (Figure 7.10). Adrenal medullary hyperplasia occurs at a lower incidence in mice compared to rats but occurs as a focal or diffuse lesion. Focal hyperplasia is

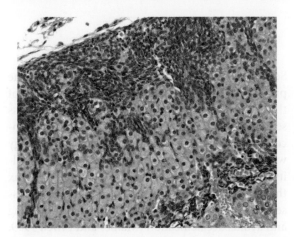

Figure 7.8 Subcapsular cell hyperplasia – type A cells predominate.

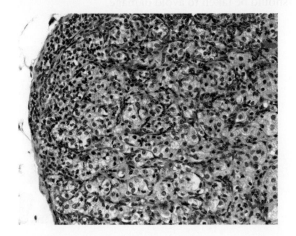

Figure 7.9 subcapsular cell hyperplasia – type B cells predominate.

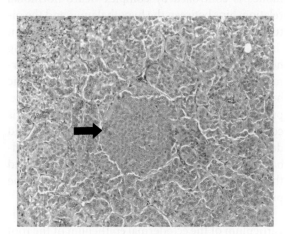

Figure 7.11 Focal medullary hyperplasia (arrow).

seen with an increased size of the affected cells with increased cytoplasmic basophilia (Figure 7.11). Diffuse changes involve an increased number and volume of all medullary cells (Nyska and Maronpot 1999; Capen *et al.* 2001).

7.3   Pituitary

7.3.1   *Background and development*

The basic structure of the mouse pituitary is the same as that of the rat and similar to other mammals, being divided into three morphologically discrete regions: the adenohypophysis, or anterior lobe (comprising the pars distalis and pars intermedia) and the neurohypophysis, or posterior lobe, which includes the pars nervosa. The different regions of the pituitary contain cells responsible for the production and release of a variety of hormones, which fall into two categories: those that act directly on target, nonendocrine organs (including growth hormone and prolactin), and trophic hormones, which modulate the activity of other endocrine organs including adrenocorticotropic hormone (ACTH) and thyroid stimulating hormone (TSH). The pituitary is functionally and anatomically connected to the brain and the production of the pituitary hormones is controlled by complex neuro-hormonal relationships between the hypothalamus and the pituitary. The production of releasing hormones from the hypothalamus (including thyrotropin-releasing hormone (TRH) and corticotropin-releasing hormone (CRH)) control the release of the trophic hormones, and the release of growth hormone, prolactin and melanocyte stimulating hormone is controlled by the production of somatostatin and dopamine from the hypothalamus. Negative feedback loops between the hypothalamus, pituitary and the target organs modulate the production of the pituitary hormones.

The pituitary gland develops from a dorsal diverticulum of the ectodermal epithelium of the roof of the oral cavity (Rathke's pouch) from embryonic days 8 to 10. The neurohypophysis derives from the infundibulum, a ventral downgrowth of neuroectoderm from the diencephalon. The wall of Rathke's pouch comes into contact with the ventral wall of the forebrain vesicle. The pars distalis and pars tuberalis originate from the anterior wall of Rathke's pouch and the pars intermedia from the portion nearest the forebrain. Hormones of the anterior pituitary can be distinguished in different cell types several days before birth. Developing nerve fibres penetrate the neural lobe anlage and additional nerve fibres enter the neural lobe as it continues to develop postnatally (Carlton and Gries 1983; Mahler and Elwell 1999).

7.3.2   *Sampling techniques*

Seymour *et al.* (2004) recommend retention of the pituitary in the skull at necropsy, and sectioning of the decalcified skull transversely to include a cross section of the pituitary. (Figure 7.12) While the process of decalcification may not affect morphology with standard H&E staining it may limit the ability to use IHC on sections. An alternative is to fix the pituitary *in situ* at necropsy, which allows dissection with minimal damage prior to embedding and avoids decalcification (Mahler and Elwell 1999; Kittel *et al.* 2004). If the gland is to be weighed, however, removal at necropsy is necessary and care should be taken to avoid damage.

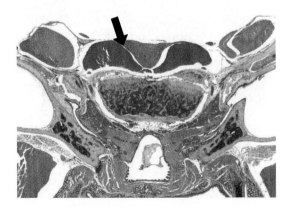

Figure 7.12   Pituitary gland *in situ* (arrow), located within the *sella turcica*.

7.3.3 Artefacts

Artefactual changes are uncommon in the pituitary gland, however, loss of the pars nervosa at trimming or sectioning may occur if the gland is not handled and embedded carefully. Apparent changes in thickness of the different components of the pituitary can result from inconsistent orientation of the gland during sectioning.

7.3.4 Anatomy and histology of the pituitary

The pituitary gland is located in the *sella turcica*, a depression in the base of the skull, lying between the trigeminal nerves, and is divided anatomically into the anterior and posterior lobes (adenohypophysis and neurohypophysis respectively). The pituitary is attached to the brain by a short stalk, which is broken upon removal of the brain, leaving the pituitary in the skull. The pituitary is larger in female than in male mice, and pregnancy is associated with a marked increase in weight of the pituitary. The major portion of the adenohypophysis is the pars distalis, which contains specific populations of endocrine cells responsible for the storage and secretion of the trophic hormones. They are arranged in cords separated by a fibrovascular stroma. These cells can be classified according to their staining characteristics in routine H&E sections, or by application of special stains such as PAS/Orange G, or the ultrastructural characteristics of the secretory granules within the cytoplasm. Based on their staining characteristics, the cells of the pars distalis are traditionally referred to as basophils, acidophils and chromophobes (Figure 7.13). Basophils include gonadotrophs that secrete luteinizing hormone (LH) and follicle stimulating hormone (FSH); thyrotrophs that release thyroid stimulating hormone (TSH); corticotrophs that secrete adrenocorticotrophic hormone (ACTH, corticotropin). Acidophils include somatotrophs, which secrete growth hormone (GH) and mammotrophs that secrete prolactin. Chromophobes do not have obvious cytoplasmic granules under

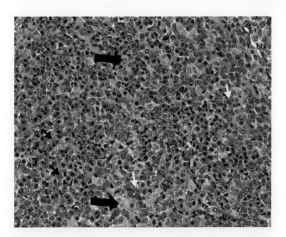

Figure 7.13 High power view of the anterior pituitary, illustrating different staining characteristics of cells of the pars distalis – acidophils (thin black arrows), basophils (white arrows) and chromophobes (thick black arrows).

light microscopy and include cells that synthesize melanocyte stimulating hormone (MSH), degranulated chromophils, and undifferentiated stem cells. Immunohistochemistry can clearly identify the contents of individual populations of cells, and it is known that some cells are able to secrete two hormones. There are differences in the distribution of different types of cells within the pars distalis. Some are diffusely distributed, whereas others are more prominent in different region of the pars distalis. For example, gonadotrophs are particularly distributed in the dorsocranial region of the pars distalis and the caudal extensions along the pars intermedia, and thyrotrophs are more evident in the ventral region. Sex differences in the populations of gonadotrophs and somatotrophs have been reported and changes in the size of prolactin secreting cells occur during the oestrous cycle and lactation in females. The pars intermedia lies between the pars distalis and the pars nervosa, separated by the residual lumen of Rathke's pouch and is relatively larger in mice than other species. It contains large polygonal, pale staining cells. The majority of ACTH and MSH is produced by the chromophobe cells of the pars intermedia in the mouse. The neurohypophysis consists of the infundibular stalk and pars nervosa.

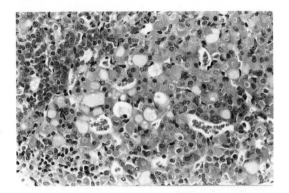

Figure 7.14   Gonadectomy cells in the pars distalis. Enlarged cells with increased eosinophilia and vacuolation.

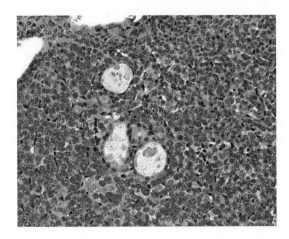

Figure 7.15   Small cysts within the pars distalis of an aged female CD-1 mouse. The cysts contain pale eosinophilic material and are lined by ciliated epithelium.

The pars nervosa appears pale and hypocellular in comparison to the pars intermedia and pars distalis and contains unmyelinated axons of hypothalamic neurons, numerous capillaries and glial cells (pituicytes). The neurohypophysis is responsible for the secretion of the hormones vasopressin, oxytocin and antidiuretic hormone (ADH). These are produced by the hypothalamic neurosecretory neurons, whose axonal processes run down the infundibular stalk and end in the pars nervosa (Mahler and Elwell 1999; Hagan *et al.* 2012).

Disruption of the negative feedback loops controlling hormone release from the pituitary, as a result of damage to, or age-related atrophy of, the target organs can lead to morphological changes in the pituitary. These can follow a characteristic pattern dependent on the population of trophic hormone-secreting cells affected. A decline in the circulating levels of hormones leads to increased stimulation of one population of endocrine cells and rapid release of their storage granules. For example, following gonadectomy, gonadotrophin secreting cells in the pars distalis undergo hypertrophic changes, with pronounced vacuolation of the cytoplasm forming characteristic 'gonadectomy cells'. Continued demand for trophic hormones can lead to focal hyperplasia (Mahler and Elwell 1999) (Figure 7.14).

One of the most common spontaneous changes in the pituitary of mice is the presence of cysts (Figure 7.15). They are commonly seen in the pars distalis or originating from Rathke's cleft, and usually are lined by ciliated epithelium and contain eosinophilic material (Mahler and Elwell 1999; Taylor 2011).

Focal hyperplasia occurs at a variable incidence in different strains of mice, usually located in the pars distalis, but occasionally affecting the pars intermedia. They form focal lesions, poorly demarcated from the surrounding parenchyma (Mahler and Elwell 1999; Taylor 2011). Hyperplasia of the pars distalis, particularly of prolactin secreting cells has been reported at a very high incidence in the FVB/N strain of mouse (Wakefield *et al.* 2003).

Pituitary tumours occur at much lower incidence in mice than in rats, but there are strain and sex differences in the incidence; they occur most commonly after 18 months of age, and their incidence is reduced by long-term dietary restriction. Female mice have a higher incidence than males and they occur more commonly in breeding females than virgins. Where they occur they usually arise from the pars distalis and can be distinguished from focal hyperplasia by the presence of a well demarcated area with distinct compression of the surrounding tissue (Mahler and Elwell 1999; Brayton 2007).

7.4   Thyroids and parathyroids

7.4.1   Background and development

The thyroid is responsible for the production and storage of iodine containing thyroid hormones (tri-iodothyronine (T_3) and tetra-iodothyronine (T_4 or thyroxine)). These hormones are responsible for regulating the basal metabolic rate and have important roles in growth and development. Thyroid hormone production is regulated by the production of thyroid stimulating hormone (TSH) from the pituitary, which is itself controlled by hypothalamic thyrotropin-releasing hormone (TRH). A negative feedback loop involving the monitoring of circulating levels of T_3 and T_4, by the hypothalamus and pituitary controls the production of the thyroid hormones. The thyroid also contains C cells, which produce calcitonin, which plays a role in calcium homeostasis. Release of calcitonin is directly regulated by circulating calcium levels. Adjacent to or embedded in the thyroids is the parathyroid gland. These are endocrine glands responsible for the production and secretion of parathyroid hormone, which works in concert with calcitonin to regulate calcium levels in the blood. Parathyroid hormone acts antagonistically to calcitonin, mobilizing reserves of calcium from the bones. Calcitonin acts to reduce serum calcium levels, by affecting osteoclast activity, calcium reabsorption from the gut and blocking renal reabsorption of calcium and phosphorus.

The thyroid is the first endocrine gland to develop in the mouse embryo, and develops from day 8 of gestation. The follicular cells originate from the first and second pharyngeal pouches and are attached to the pharyngeal pouch by the thyroglossal duct. C cells develop from ultimobranchial bodies, originating from the neural crest, which fuse with the thyroid primordium. By gestation day 15, small buds are evident, and the parathyroid gland and C cell precursors are embedded within the thyroid, but follicle development does not start until day 17. Colloid filled follicles are evident by day 18 and active secretory activity begins on day 19. Parathyroids develop from the third and fourth pharyngeal pouch and are closely associated with the thymus during embryonic development. They are visible from around gestation day 11.5 (Hardisty and Boorman 1999; La Perle and Jordan 2012).

7.4.2   Sampling techniques

The thyroids can be retained *in situ*, if not required to be weighed, and trimmed transversely still attached to the trachea, presenting a cross section of trachea, oesophagus, thyroid and parathyroid together (Figures 7.16 and 7.17). Alternatively, if the thyroids have to be weighed, they can be removed from the trachea and embedded whole to present a longitudinal section with the parathyroid (Kittel *et al.* 2004).

7.4.3   Artefacts

Care needs to be taken when removing the thyroids from the trachea. If thyroid weights are required, it may be more suitable to weigh the tissues, post-fixation (after approximately 24 hours) to reduce the chances of damaging the tissue. Crush artefacts

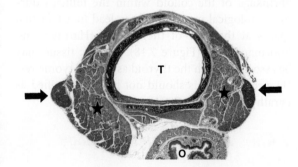

Figure 7.16   Sample of thyroid (stars) and parathyroid (arrows) in cross section, including trachea (T) and oesophagus (0).

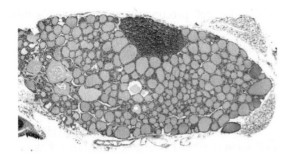

Figure 7.17 Sample of thyroid and parathyroid in longitudinal section following removal from trachea.

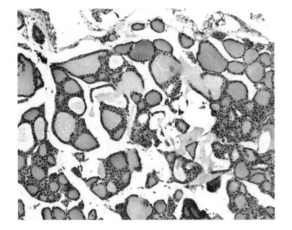

Figure 7.18 Thyroid gland, illustrating artefactual crush damage.

are occasionally seen on the periphery of the thyroid gland as a result of inappropriate handling at necropsy or histology (La Perle and Jordan 2012). Shrinkage of the colloid within the follicles during histological processing, may lead to artefactual spaces at the edge of the follicles (Hardisty and Boorman 1999) (Figure 7.18). Thymic tissue may be associated with the thyroid as an embryonic rest (Figure 7.22). This should not be misidentified as parathyroid (Krinke 2004).

7.4.4 Anatomy and histology of the thyroids

The paired thyroid glands are located on either side of the trachea, near the base of the laryngeal cartilages. A thin isthmus may be apparent between the two glands. The glands are covered by a thin fibrous capsule, and fibrous septae from this capsule extend into the gland. Position and number of parathyroid glands can vary; there are usually two parathyroids in the mouse, located under the capsule. The thyroid contains multiple follicular structures, lined by a single layer of epithelium, with the lumen containing variable amounts of eosinophilic colloid. The height of the follicular epithelium and the size of the colloid-filled follicles is dependent on the metabolic activity of the animal, and can vary considerably within the gland itself (Figure 7.19). Thyroid gland morphology is influenced by a variety of nutritional, environmental and endocrine factors as well as showing sex and strain differences. Metabolically active follicles have a high cuboidal follicular epithelium, and the follicles are relatively small. As the animal ages, however, and metabolic activity decreases, an increased number of so-called 'cold' follicles become apparent (Figure 7.20). These large follicles have a flattened epithelium, and increased amounts of pale-staining colloid. The total number of follicles within the thyroid remains the same during the lifetime of the mouse. The increased size and weight of the thyroids reported in ageing mice is attributed to the increased presence of large 'cold' follicles. A few C cells are also present in the mouse thyroid, although they are not usually apparent in routine H&E-stained sections. Immunohistochemical

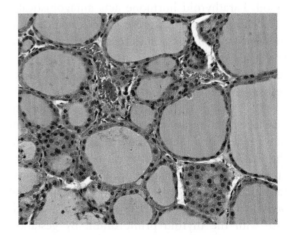

Figure 7.19 Thyroid gland, illustrating variation in follicular size in a control mouse.

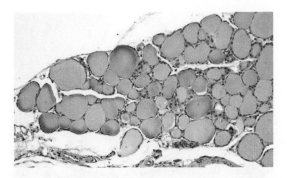

Figure 7.20 Thyroid gland of aged mouse illustrating presence of 'cold' follicles. Enlarged follicles lined with flattened epithelium.

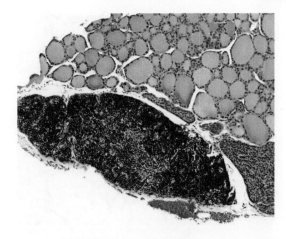

Figure 7.22 Thyroid – ectopic thymus.

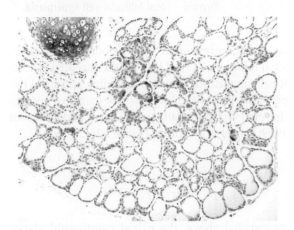

Figure 7.21 Thyroid gland of mouse illustrating distribution of calcitonin positive C cells labeled using IHC (photo courtesy of Andrew Pilling).

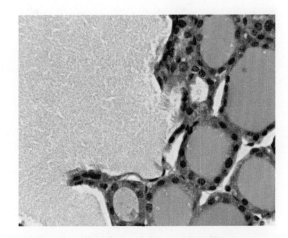

Figure 7.23 High power view of ultimobranchial cyst illustrating the presence of ciliated epithelium.

staining for calcitonin can reveal the cells, usually located near the centre of the gland (Figure 7.21). The C cells are located at the basal region of follicular cells, within the thyroid follicles (Hardisty and Boorman 1999; La Perle and Jordan 2012).

Ectopic tissues and embryonic rests are occasionally found associated with the thyroid. Ectopic thyroid is reported at the base of the heart (Frith and Ward 1988) and ectopic thymus tissue is regularly seen associated with the thyroids (Taylor 2011) (Figure 7.22). Cystic remnants of embryonic tissues are common background findings in the mouse, with a very high incidence in some strains

(Figure 7.23). They occur in or around the thyroid and can be lined by ciliated, cuboidal epithelia or squamous epithelial cells and are considered to be remnants of the craniopharyngeal duct or ultimobranchial duct (Frith *et al.* 2007; Taylor 2011).

Cystic dilatation of follicles is occasionally seen with marked enlargement of colloid containing follicles (Figure 7.24). Crystals are occasionally seen within the colloid of the mouse thyroid that show birefringence under polarized light (Figure 7.25). The crystals have been identified as calcium oxalate in humans (Taylor 2011).

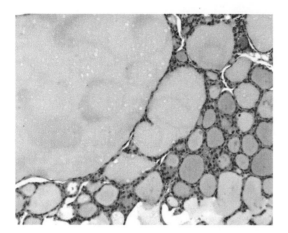

Figure 7.24 Thyroid – cystic dilatation of follicles.

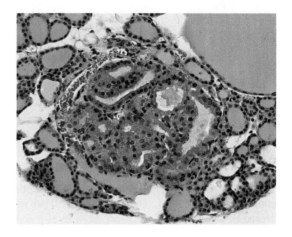

Figure 7.26 Thyroid – Focal follicular cell hyperplasia.

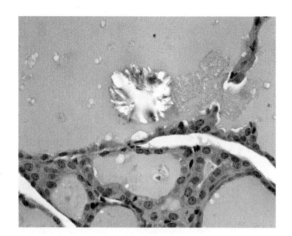

Figure 7.25 Thyroid – crystals within follicular colloid.

The thyroids are described as a common site for deposition of amyloid in different strains of mice, particularly the CD-1 mouse (Brayton 2007). However, the occurrence of amyloidosis in CD-1 mice has shown time-related changes and can vary considerably between animal suppliers (Maita *et al.*1988; Engelhardt *et al.* 1993; Taylor 2011).

Hyperplastic changes and tumours of the thyroid follicular cells and C cells are relatively uncommon compared to the rat. Follicular and C cell proliferative lesions occur as a continuum from hyperplasia to adenoma and carcinoma and distinction between the three is based on morphology and behaviour. Diagnosis of hyperplasia is based on a lack of compression around a poorly demarcated, focal lesion. For C cell lesions, the accepted criteria for distinguishing between adenoma and focal hyperplasia is that the lesion is less than five average thyroid follicles in diameter (Hardisty and Boorman 1999; Capen *et al.* 2001) (Figure 7.26).

7.4.5 Anatomy and histology of the parathyroids

As detailed above, the paired parathyroid glands (usually two in the mouse) are within the thyroid capsule and are closely associated with or embedded within the thyroid tissue. They are usually not sampled separately as identification of the parathyroids at necropsy is difficult. The presence of pigmented melanocytes in the stroma of the parathyroid in pigmented strains of mice may aid identification of the parathyroids at necropsy (Russfield 1967). There is a thin capsule of connective tissue separating the parathyroid from the thyroid. The parathyroid is composed of a single population of polygonal cells with eosinophilic cytoplasm, the chief cells, which are arranged in cords with a fine fibrovascular stroma. Oxyphil cells are not present in the mouse (Capen *et al.* 1996).

Ectopic parathyroid tissue may be found associated with the lobular septae of the thymus, and thymic tissue may be associated with the

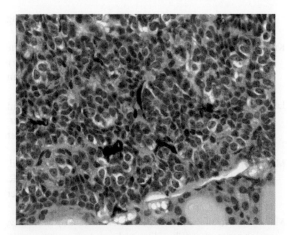

Figure 7.27   Parathyroid – melanocytes in the stroma of a pigmented mouse.

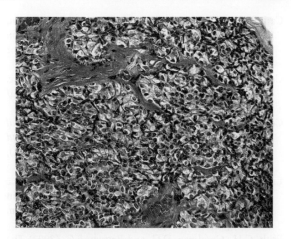

Figure 7.29   Parathyroid – increased interstitial fibrous tissue in an aged mouse.

parathyroid as a result of their similar embryonic origin (Hardisty and Boorman 1999). Melanocytes are seen in the stroma of the parathyroid in pigmented strains of mice (Figure 7.27).

Parathyroid cysts are occasionally seen, and are usually lined by a ciliated epithelium and contain eosinophilic material (Capen *et al.* 1996; Hardisty and Boorman 1999) (Figure 7.28). An increase in interstitial fibrous connective tissue is often seen as an age-related change in the parathyroid (Anver and Haines 2004) (Figure 7.29).

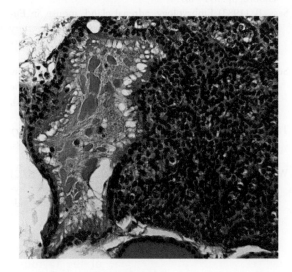

Figure 7.28   Parathyroid cyst.

References

Anver, M.R. and Haines, D.C. (2004) Gerontology, in *The Handbook of Experimental Animals: The Laboratory Mouse* (eds H.J. Hedrich and G. Bullock), Elsevier, San Diego, CA, pp. 327–343.

Bernichtein, S., Peltoketo, H., Huhtaniemi, I. (2009) Adrenal hyperplasia and tumors in mice in connection with aberrant pituitary-gonadal function. *Molecular and Cellular Endocrinology* **300**, 164–168.

Bielinska, M., Kiiveri, S., Parviainen, H. *et al.* (2006) Gonadectomy-induced adrenocortical neoplasia in the domestic ferret (*Mustela putorius furo*) and laboratory mouse. *Veterinary Pathology* **43**, 97–117.

Bielohuby, M., Herbach, N., Wanke, R., *et al.* (2007) Growth analysis of the mouse adrenal gland from weaning to adulthood: time- and gender-dependent alterations of cell size and number in the cortical compartment. *American Journal of Physiology – Endocrinology and Metabolism* **293**, E139–E146.

Brayton, C. (2007) Spontaneous diseases in commonly used mouse strains, in *The Mouse in Biomedical Research*, 2nd edn, Vol II, Diseases (eds J.G. Fox, S.W. Barthold, M.T. Davisson *et al.*), Academic Press, Burlington, MA, pp. 623–717.

Capen, C.C., Gröne, A., Bucci, T.J. *et al.* (1996) Changes in structure and function of the parathyroid gland, in *Pathobiology of the Aging Mouse*, Vol 1 (eds U. Mohr, D.L. Dungworth, C.C. Capen *et al.*), ILSI Press, Washington, D.C., pp. 109–123.

Capen, C.C., Karbe, E., Deschl, U., *et al.* (2001) Endocrine system, in *International Classification of Rodent Tumors, The Mouse* (ed. U. Mohr), Springer, Berlin, pp. 269–322.

Carlton, W.W. and Gries, C.L. (1983) Cysts, pituitary: rat, mouse, and hamster, in *Endocrine System, Monographs on Pathology of Laboratory Animals* (eds T.C. Jones, U. Mohr and R.D. Hunt), Springer-Verlag, Berlin, pp. 161–163.

Cook, M.J. (1965) *The Anatomy of the Laboratory Mouse*, Academic Press, London, http://www.informatics.jax.org/cookbook/ (accessed 18 July 2013).

Cook, M.J. (1983) Anatomy, in *The Mouse in Biomedical Research Vol III, Normative Biology, Immunology, and Husbandry* (eds H.L. Foster, J.D. Small and J.G. Fox), Academic Press Inc., Boston, MA, pp. 101–120.

Engelhardt, J.A., Gries, C.L. and Long, G.G. (1993) Incidence of spontaneous neoplastic and nonneoplastic lesions in Charles River CD-1 mice varies with breeding origin. *Toxicology and Pathology* **21**, 538–541.

Frith, C.H., Goodman, D.G. and Boysen, B.G. (2007) The mouse, pathology, in *Animal Models in Toxicology*, 2nd edn (ed. S.C. Gad), Taylor & Francis, Boca Raton, FL, pp. 72–122.

Frith, C.H. and Ward, J.M. (1988) *Color Atlas of Neoplastic and Non-Neoplastic Lesions in Ageing Mice*, Elsevier, Amsterdam, www.informatics.jax.org/frithbook/index.shtml (accessed 15 August 2013).

Hagan, C.E., Bolon, B. and Keene, C.D. (2012) Nervous system, in *Comparative Anatomy and Histology: A Mouse and Human Atlas* (eds P.M. Treuting and S.M. Dintzis), Elsevier Inc., Amsterdam, pp. 339–394.

Hardisty, J.F. and Boorman, G.A. (1999) The thyroid and parathyroid glands, in *Pathology of the Mouse, Reference and Atlas* (eds R.R. Maronpot, G.A. Boorman and B.W. Gaul), Cache River Press, Vienna, pp. 537–554.

Hummel, K.P., Richardson, F.L. and Fekete, E. (1966) Anatomy, in *Biology of the Laboratory Mouse* (ed. E.L. Green), McGraw-Hill, New York, pp. 247–307, http://www.informatics.jax.org/greenbook/frames/frame13.shtml (accessed 18 July 2013).

Kim, J.S., Kubota, H., Nam, S.Y. *et al.* (2000) Expression of cytokines and proteases in mast cells in the lesion of subcapsular cell hyperplasia in mouse adrenal glands. *Toxicologic Pathology* **28**, 297–303.

Kittel, B., Ruehl-Fehlert, C., Morawietz, G. *et al.* (2004) Revised guides for organ sampling and trimming in rats and mice – Part 2. *Experimental Toxicologic Pathology* **55**, 413–431. Available electronically at http://reni.item.fraunhofer.de/reni/trimming/index.php (accessed 18 July 2013).

Knoblaugh, S., Randolph-Habecker, J. and Rath, S. (2012) Necropsy and histology, in *Comparative Anatomy and*

Histology: A Mouse and Human Atlas (eds P.M. Treuting and S.M. Dintzis), Elsevier Inc., Amsterdam, pp. 15–40.

Krinke, G.J. (2004) Normative histology of organs, in *The Handbook of Experimental Animals: The Laboratory Mouse* (eds H.J. Hedrich and G. Bullock), Elsevier, San Diego, CA, pp. 133–166.

La Perle, K.M. and Jordan, C.D. (2012) Endocrine System, in *Comparative Anatomy and Histology: A Mouse and Human Atlas* (eds P.M. Treuting and S.M. Dintzis), Elsevier Inc., Amsterdam, pp. 211–227.

Mahler, J.F. and Elwell, M.R. (1999) The pituitary gland, in *Pathology of the Mouse, Reference and Atlas.* (eds R.R. Maronpot, G.A. Boorman and B.W. Gaul), Cache River Press, Vienna, pp. 491–507.

Maita, K., Hirano, M., Harada, T. *et al.* (1988) Mortality, major cause of moribundity, and spontaneous tumors in CD-1 mice. *Toxicologic Pathology* **16**, 340–349.

Nyska, A. and Maronpot, R.R. (1999) The adrenal gland, in *Pathology of the Mouse, Reference and Atlas* (eds R.R. Maronpot, G.A. Boorman and B.W. Gaul), Cache River Press, Vienna, pp. 509–536.

Rosol, T.J., Yarrington, J.T., Latendresse, J. and Capen, C.C. (2001) Adrenal gland: structure, function, and mechanisms of toxicity. *Toxicologic Pathology* **29**, 41–48.

Ruehl-Fehlert, C., Kittel, B., Morawietz, G. *et al.* (2003) Revised guides for organ sampling and trimming in rats and mice – Part 1. *Experimental Toxicologic Pathology* **55**, 91–106, http://reni.item.fraunhofer.de/reni/trimming/index.php (accessed 18 July 2013).

Russfield, A.B. (1967) Pathology of the endocrine glands, ovary and testis of rats and mice, in *Pathology of Laboratory Rats and Mice* (eds E. Cotchin and F.J.C. Roe), Blackwell, Oxford, pp. 391–467.

Sass, B. (1983) Embryology, adrenal gland, mouse, in *Endocrine System, Monographs on Pathology of Laboratory Animals* (eds T.C. Jones, U. Mohr, R.D. Hunt), Springer-Verlag, Berlin, pp. 3–7.

Seymour, R., Ichiki, T., Mikaelian, I. *et al.* (2004) Necropsy methods, in *The Handbook of Experimental Animals: The Laboratory Mouse* (eds H.J. Hedrich and G. Bullock), Elsevier, San Diego, pp. 495–516.

Taylor, I. (2011) Mouse, in *Background Lesions in Laboratory Animals. A Color Atlas* (ed. E.F. McInnes), Saunders, Elsevier, Edinburgh, pp. 45–72.

Wakefield, L.M., Thordarson, G., Nieto, A.I. *et al.* (2003) Spontaneous pituitary abnormalities and mammary hyperplasia in FVB/NCr mice: implications for mouse modeling. *Complementary Medicine* **53**, 424–432.

Chapter 8

Nervous system

Aude Roulois
GlaxoSmithKline R&D, UK

8.1   Introduction

The nervous system of the mouse is similar to that of other mammals in being divided anatomically and functionally into the central nervous system (CNS), consisting of the brain and spinal cord, and the peripheral nervous system (PNS), which connects the CNS to other organs and includes motor and sensory cranial and peripheral nerves and autonomic ganglia (controlling visceral functions).

With the widespread use of mice as biological models, an extensive literature has developed on the anatomy, physiology, development and pathology of its nervous system. Many resources are available online (see references to web sites at the end of this chapter).

8.2   Necropsy

The brain and spinal cord are lipid-rich, soft, fragile tissues, and therefore prone to microscopic and macroscopic artefacts. Prevention or limitation of artefacts during necropsy and subsequent tissue preparation is widely discussed in the literature because of the practical challenges in distinguishing between artefacts and pathological lesions. The principles used to reduce artefact are the same for all species and numerous general reviews of techniques are available that can be used in developing best practices for nervous tissue sampling in mice (Garman 2011; Spijker 2011; Bolon *et al.* 2013; Jordan *et al.* 2011).

Cervical dislocation can lead to extensive haemorrhage and traumatic damage especially to the hindbrain and cranial spinal cord and should, if possible, be avoided as a method of euthanasia when examination of the CNS is required. It is preferable to sever the head from the rest of the body after death with a sharp scalpel blade introduced at the level of the first cervical vertebra before dissecting the brain out of the skull (Figure 8.1). After this, although details in the necropsy techniques vary to some extent between laboratories, reflecting local experience

A Practical Guide to the Histology of the Mouse, First Edition. Cheryl L. Scudamore.
© 2014 John Wiley & Sons, Ltd. Illustrations © Veterinary Path Illustrations, unless stated otherwise.
Published 2014 by John Wiley & Sons, Ltd. Companion Website: www.wiley.com/go/scudamore/mousehistology

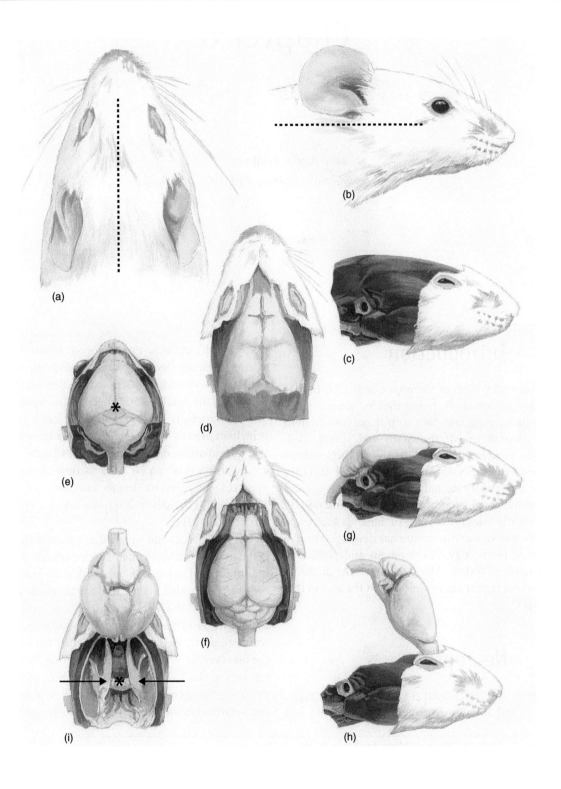

(a)

(b)

(c)

(d)

(e)

(f)

(g)

(h)

(i)

Figure 8.1   Necropsy. A dorsal longitudinal section (line in (a)) is carried with a scalpel blade from the nose down to the neck. Dorsal and lateral skin, subcutaneous tissues and muscles are removed to below the eyes and ears level (line in (b) and (c)). The skull is then severed from the rest of the body at the level of the first cervical vertebra with a sharp scalpel (to avoid crushing the enclosed cord). At this stage, the brain can be seen through the skull cap (i.e. the calvarium) (d) which can then be removed (e). For this, fine scissors are inserted into the foramen magnum and the skull cap is progressively lifted with fine forceps. The position of the pineal gland (which may remain attached on the skull cap) is indicated (asterisk in (e)). After removing the skull cap, and to avoid damaging the brain when lifting it out of the skull, care should be taken in preliminarily removing the cerebellar tentorium, transversally separating the cerebral cortex from the cerebellum and the cerebral falx, longitudinally separating both cerebral hemispheres. Once the skull cap and meninges are removed, the brain can be immersion fixed in situ or dissected further out of the skull. In this latter case, freeing the cranial end of the brain requires a sharp full thickness transversal cut through the olfactory bulbs (if their examination is not required) or further cranially (if the nasal cavities are not being preserved) removing sufficient bone on each sides to ensure that the brain (with or without remaining olfactory bulbs) can easily be lifted out of the skull without being ripped (dorsal view (f) and lateral view (g)). Finally, the ventral cranial nerves and pituitary are carefully sectioned to free the brain from the base of the skull (h). The pituitary (* in (i)) which is located in a small recess (the sella), can be separately fixed in situ with the remaining ventral skull (i). Likewise, the trigeminal ganglia (arrows), which sit on the ventral surface of the skull with its afferent nerves running cranially and efferent roots caudally, can be fixed or further isolated by freeing them from the thick covering dura.

←

and preference, examination of the CNS typically requires opening the bone in which the tissue is enclosed (Figure 8.1, Chapter 1, Figure 1.50). Once the head is severed from the body, the easiest way to remove the skull cap is to proceed cranially from the foramen magnum (the caudal opening in the skull from which the spinal cord protrudes). How far cranially the skull cap is removed will determine whether the olfactory bulbs or the nasal cavities are preserved for macroscopic and microscopic examination as it can be difficult to preserve both optimally. For this reason, it is worth considering ahead of time if the examination of the olfactory bulbs, which are an important region, rich in neural progenitors, will be required. Variable sampling of this region is also a source of inter-animal variation when brain weights are collected.

Of the three layers of meninges separating the skull from the brain (or the vertebral body from the spinal cord) the dura mater is the most external and the toughest. Going inwards, are the richly vascularized pia arachnoid and the pia mater. Individual layers are not readily distinguishable at necropsy, but the meninges need to be removed before the brain can be extracted from the skull (Spijker 2011). This is best done by grasping them with a pair of forceps at the level of their two fibrous extensions, the cerebellar tentorium and the cerebral falx

(Figure 8.1). Once removed, the naturally smooth surface of the brain is revealed (the mouse is a lissencephalic species lacking the gyri and sulci seen in the brain of gyrencephalic species) (Figure 8.1). The skull, containing the brain, can then be tilted upside down and the cranial nerves and pituitary cut ventrally while preventing pulling and stretching of the tissue as it is freed from the bone (Figure 8.1). Once the brain is removed, the floor of skull will still contain the pituitary in its sella, together with the trigeminal ganglia and nerves (Figure 8.1). The pineal gland that sits dorsally at the junction between the cerebral hemispheres and the cerebellum can be left attached to the skull cap.

The vertebral column, containing the spinal cord, is revealed by sectioning the skin along the back of the animal and removing the dorsal muscles on each side. The vertebral column can then be separated from the rest of the body by cutting along the proximal end of the ribs and the base of the sacrum. From there, the column can be either partially opened (laminectomy) or processed with the cord in situ (see section 8.3).

Routine examination of the PNS in mice is usually restricted to examination of a unilateral section of sciatic nerve (peripheral nerve). The sciatic nerves run along the lateral-caudal side of the femur and are easily accessed dorsally by sectioning

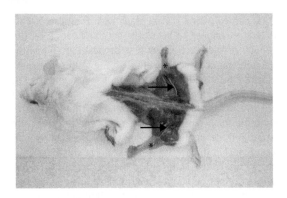

Figure 8.2   Sciatic nerve necropsy. A dorsal longitudinal dissection has been performed with a scalpel blade along the midline revealing the vertebral column. The dorsal muscular bundles (black asterisk) are severed from the column and pelvis with a scalpel to expose the white sciatic nerve (arrows) on each side.

the thigh muscles along the sacrum and spreading them laterally (Figure 8.2).

At this stage, it is helpful to preserve the proximal / distal orientation of the spinal cord and nerves. This can be achieved by cutting the proximal and distal end of the cord / nerve at a different angle (straight cut at one end, oblique at the other), by tying a fine thread of material at one end of the sample (Krinke *et al.* 2000b) or by adhering the sample to a cardboard, the orientation being labelled with a pencil (ink can be washed out by solvents). Preserving the samples from stretch or crush can be achieved by manipulating them with care, holding them by their extremities, by using sharp blade or scissors for sectioning them and by laying them flat on a piece of cardboard before dropping them into the fixative. Techniques for preparing peripheral nerves for immersion fixation have been well described elsewhere (Jortner 2011).

Sampling of other nerves (cranial or peripheral) and ganglia is generally restricted to studies where they are of specific interest and are not described in detail here but protocols can be found in the literature (Shimeld *et al.* 2001; Bolon *et al.* 2006 and 2008).

Macroscopic observations visible at necropsy may be lost once the tissues have been further dissected or fixed and so it is important that these are described at necropsy (see Chapter 2, Table 2.3

for example descriptors). Here, an awareness of the visible anatomical landmarks helps to distinguish subtle changes (Figure 8.3). For instance, a proportional decrease in the size of the cerebellum, when compared with the cerebral cortex and medulla oblongata adjacent to it, can be used to diagnose cerebellar atrophy.

Neuroanatomical terminology used in describing the CNS can be confusing, as some is based on the macroscopic landmarks, the developmental origin or on its functional units. Confusion and overlap may inevitably occur, which are also the reflection of our evolving knowledge of the CNS, of its functions and of the comparative similarity (or difference) between the mouse and other species. (For a comparative review of mouse and human brains the reader is referred to Hagan *et al.* (2011)). There is also potential for confusion over some commonly used terms. "Brainstem", for instance, can be used to name the pons and medulla oblongata, but sometimes additionally includes parts of the midbrain and diencephalon. For descriptive purposes, lesions can be located using anatomical landmarks (Figure 8.3) complemented with qualifiers (for example, red focus on the cranioventral surface of the left olfactory bulb). Familiarity with these descriptive landmarks is therefore helpful to locate the changes observed (and their possible relationship to clinical signs). Being consistent and specific to be understood by others and allow data comparison across observers and/or studies is also valuable. It may not always be necessary to link the location of a lesion to an individual nucleus but using 'brain' as a lesion locator can be insufficient. In practice, a template of useful locators for recording macroscopic and microscopic lesions (either as a tabulated list of structures to examine or as a diagram) prior to starting the work can be time well spent (Figure 8.3 and Table 8.1).

8.3   Fixation/perfusion

The susceptibility of the nervous tissue to artefact also impact on its fixation. Historical habit and the prominence of specialist neuropathology techniques in the literature have led to the frequent

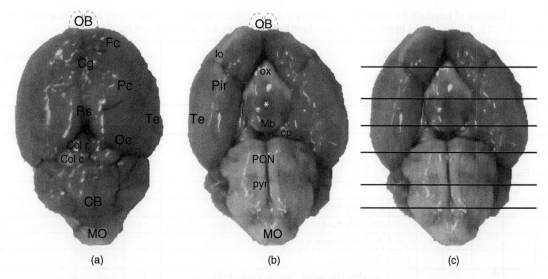

Figure 8.3 Macroscopic observation of the mouse brain. (a) On the dorsal surface of the brain (from rostral to caudal) one can recognize the position of the two olfactory bulbs (OB) and smooth cerebral hemispheres of the cerebral cortex which divide into distinct regions: the frontal (Fc), parietal (Pc), occipital (Oc), temporal (Te), cingulate (Cg), retrosplenial (Rs) and, on the ventral surface, the piriform (Pir) cortex. Also on the dorsal surface of the brain are the emplacement of the pineal gland (black asterisk) left attached on the skull cap, the superior and inferior colliculi -respectively referred to as anterior or rostral (Col r) and posterior or caudal (Col c) when respecting the spatial orientation in a quadruped mouse, the cerebellum (CB)-divided into a central vermis and two lateral hemispheres and extending at its widest part into the paraflocculi- and the medulla oblongata (MO) beneath it. (b) Rostro-caudal examination of the ventral surface of the mouse brain shows again the olfactory bulbs, the cortical hemispheres on which may sit the two optic nerves converging caudally into the optic chiasm (ox), the ventral surface of the hypothalamus which surrounds the infundibulum (white asterisk) where the pituitary attaches to the brain, the mamillary bodies (Mb), cerebral peduncles (cp) and the pons (PON) in continuity with the medulla oblongata along which run the pyramids (pyr). (c) These ventral macroscopic landmarks can be used to select the transversal sections prepared for histology (the black lines indicate approximate positions of the transversal sections presented on Figure 8.11).

recommendation to use perfusion fixation for CNS and/or PNS examination.

Perfusion fixation prior to tissue handling has advantages because, providing it is correctly done, it minimizes the time between death and tissue fixation. The fixative is directly infused through the vasculature antemortem under deep anaesthesia, minimizing autolysis and some of the perimortem artefacts induced by hypoxia, tissue handling or processing (mainly dark neurons and vacuolation). Perivascular retraction of the neuropil, which can be seen in immersion fixed tissues and may be worsened with hyperosmolar fixatives (Fix and Garman 2000), is another artefact not seen in well-perfused fixed tissues, unless too great a pressure is applied in the perfusion system. Perfusion fixation can also reveal subtle microscopic changes, which are not

readily visible in immersion fixed tissues (Auer and Coulter 1994).

However, it has disadvantages in that it is time-consuming, potentially hazardous for the operator (inhalation of fixative fumes) and also limits the flexibility of using different fixatives for different tissues from the same animal at the time of necropsy. It also requires additional equipment, careful planning, proficiency of the operator in the technique and familiarity for the pathologist subsequently examining the tissue microscopically (as the tissue morphology will differ slightly from that of immersion fixed tissue: for example blood vessels will appear patent and empty) (Figure 8.4). Additionally, suboptimal perfusion may not deliver the expected advantages over immersion fixation and may increase interanimal variability in tissue morphology, which may in

Table 8.1 Example of template for recording macroscopic and microscopic lesions in the mouse brain. Adapted from Bolon *et al.* 2013. The structures listed here can be identified on figures 8.3, 8.8, 8.11 and/or 8.12.

							Examined		Not examined/ not present
							NAD	Other	
Brain									
Structure present on section (from Figure 8.11)									
a	b	c	d	e	f				
						Olfactory bulb			
✓						Olfactory tract			
✓						Anterior commissure			
✓						Septal nuclei			
✓						Striatum (including caudate / putamen)			
✓	✓	✓	✓			Cerebral cortex (including frontal, parietal, temporal, occipital and piriform areas)			
	✓	✓				Amygdala			
✓	✓					Corpus callosum			
	✓					Internal capsule			
✓	✓	✓	✓			External capsule			
✓						Optic tracts / optic chiasm			
	✓	✓				Mamillothalamic tract			
	✓					Fasiculus retroflexus			
			✓			Medial lemniscus			
	✓	✓				Fimbria/fornix			
	✓	✓				Hippocampus (including Ammon's horns an dentate gyrus)			
	✓	✓				Thalamus and geniculate bodies			
	✓					Hypothalamus			
	✓	✓	✓			Cerebral peduncles			
		✓	✓			Midbrain (including substantia nigra, colliculi an periaqueductal grey matter)			
			✓			Pons			
				✓	✓	Pyramids			
				✓	✓	Cerebellum			
				✓		Deep cerebellar nucleis			
				✓	✓	Posterior cerebellar peduncles			
				✓		Reticular formation			
			✓	✓	✓	Raphe			
				✓	✓	Trigeminal nuclei and tracts			
				✓	✓	Medulla oblongata (including cochlear facial, vestibular, trigeminal and olivary nuclei)			
	✓				✓	Choroid plexus			

Table 8.1 (*continued*)

	Examined		Not examined/ not present
	NAD	**Other**	
Subventricular organs (including median eminence, subfornical and subcomissural organs) (see figure 8.12)			
Ventricular system (including lateral, 3d, 4th ventricles and aqueduct of Sylvius)			
Spinal cord			
Cervical segment (grey and white matter)			
Thoracic segment (grey and white matter)			
Lumbar segment (grey and white matter)			
Peripheral nerve(s)			
Sciatic (unilateral)			
Other(s), specify			
Others			
Eye and optic nerve			
Other special sense organ(s), specify			
Muscle, specify			

turn compromise the comparative microscopic examination. For these reasons and given that an experienced pathologist can, in the majority of cases, differentiate pathological lesions from those artefactual changes present in immersion fixed tissue, this simpler technique is considered acceptable when examination of the nervous tissue forms part of a routine screen of all organs rather than a specific study of the nervous system (Butt 2011).

Although 4% PFA is often chosen in the field of neurosciences for IHC and specialized neuropathological examination on the basis that alcohol contained in ready-made NBF solutions (and in other alcohol based fixatives) can cause artefactual vacuolation (Figure 8.4), 10% NBF may be suitable for many routine examinations. Another commonly used fixative in neurosciences is glutaraldehyde (either alone at 4% or in combination with PFA), which provides optimal fixation for EM. Specific examination of myelin sheaths by electron microscopy is a good example of when the combined use of glutaraldehyde and perfusion are

desirable for optimal preservation of morphology (poorly fixed myelin sheath show artefactual separation –from the axon and / or between consecutive myelin layers – vacuolation and loss of cylindrical shape). Osmium fixation or postfixation is also used in some protocols. This provides increased contrast suitable for EM examination, but is not necessary for routine H&E pathological examination.

Freeze fixation, alone or in combination with preliminary perfusion, is best done by immersion of the tissue in cooled isopentane (sometimes preceded by immersion in a hypertonic sucrose solution) rather than by direct contact with dry ice or liquid nitrogen to protect cell membranes and tissue morphology. Such protocols, suitable for preserving genetic material (for *in situ* hybridization or PCR) and / or some proteins (for the IHC of some antigens), are not recommended for morphological examination of FFPE H&E sections, as immersion in hypertonic solutions and freezing induce tissue shrinkage and freeze artefacts.

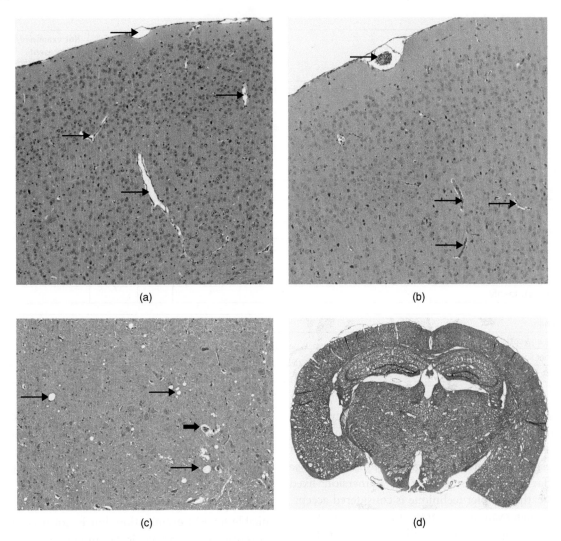

(a)   (b)

(c)   (d)

Figure 8.4   Perfusion or immersion fixations and artefacts. (a) Blood vessels are patent and devoid of red blood cells in perfused nervous tissue (a section of cerebral cortex). (b) In contrast in immersion fixed tissue (in a comparable section of cerebral cortex), blood vessels are collapsed and/or contain red blood cells (arrows). (c) Artefactual vacuolation (blue arrows) can be seen in immersion fixed tissue, often located in the deep cerebellar nuclei/peduncles region. These vacuoles are variably sized, often empty with smooth rounded edges. The adjacent neurons appear morphologically normal and the cellularity of the area is unremarkable (indicating the absence of any glial reaction associated with the presence of the vacuoles). Perivascular retraction of the neuropil (an artefact of immersion fixation) can also be seen here (thick arrow). (d) Excessive pressure applied in a perfusion system can lead to severe artefactual vacuolation of the tissue giving it a 'Swiss cheese' appearance, not unlike that of severe freezing artefact.

The spinal cord can be fixed and processed in situ with the vertebrae (Figure 8.5). However, decalcification of the bone (to soften the tissue so that it can be cut on the microtome) prior to tissue processing is often done using formic acid. This has a negative impact on morphology (Knoblaugh *et al.* 2011) and results in shrunken cell nuclei with poor cytoplasmic and nuclear details (Figure 8.5). Partial dissection (laminectomy of a small segment of the column) or complete dissection of the spinal cord from the vertebral column prior to fixation can improve morphology but this has to be weighed against the time it takes, the delay in the time to fixation and the potential secondary artefacts (autolysis, crush, stretch etc.). Complete dissection of the cord from the vertebrae may also result in loss of the dorsal root ganglia (DRGs). For these reasons, perfusion fixation may be required when optimal morphological preservation of the cord, ganglia or nerves is needed for detailed and specific microscopic examination.

8.4   Trimming

The value of careful examination of macroscopic landmarks on the surface of the brain and of their relative size and positions is mentioned in section 8.2, with alteration in size, colour, shape and relative positions potentially representing part of a phenotype (congenital/developmental or acquired). These landmarks can also be used to assist the sampling / slicing of the brain (Radovsky and Mahler 1999; Spijker 2011). Sectioning reveals new structures, which can also be examined for the presence of abnormalities, before the tissue is processed.

A brain atlas (hard copy, electronic or web-based versions, see references) can be useful to learn and identify anatomical structures (Paxinos and Franklin 2012). Familiarity with the neuroanatomy of the spinal cord, for which reviews and atlases are available, can also help integrating morphological lesions with clinical signs (Bolon 2000).

Most brain atlases provide stereotactic coordinates to reference the position of each brain (or spinal cord) structure on a 3D grid. These coordinates are used for targeted neurosurgery using a stereotactic frame fixed on the animal's head (or back). These atlases use a reference point called bregma, identified by the intersection of the coronal and sagittal sutures on the skull at the junction between frontal and parietal bones. As an alternative to macroscopic landmarks, brain trimming can also be done using a matrix (also known as a brain slicer). These matrices (Figure 8.6) are available for a range of ages and sizes of mouse brain, to produce standard sections of a range of thicknesses and orientations. They potentially allow more consistent sectioning between animals and the production of sections matching that of an atlas (symmetrical along the median and horizontal axis, which makes the comparison of the microscopic section to that of the atlas easier). In practice, the advantages need to be balanced against the cost of these matrices and the degree of variability that still can be introduced depending on how perfectly the brain is positioned in the mould and the proficiency of the technician. In particular, how far caudally the medulla is cut and whether the olfactory bulbs are present will influence the position of the brain in the mould.

Trimming guidelines using macroscopic landmarks or stereotactic coordinates are available to assist developing a trimming scheme for the mouse brain. Most are derived from the field of toxicological pathology (Radovsky and Mahler 1999; Morawietz *et al.* 2004; Jordan *et al.* 2011; Bolon *et al.* 2013). Although often not specifically designed for mice (describing general procedures for rodents or specific to rats), the principles and types of sections they describe are relevant. What represents an adequate sample of nervous tissue (and of brain in particular) for routine microscopic evaluation is the subject of debate (Rao *et al.* 2011; Switzer *et al.* 2011; Bolon *et al.* 2013) centred on which anatomical and/or functional regions need to be examined given the relative complexity (anatomically and functionally) of the nervous system, the discrepancy that may occur between clinical observations and microscopic morphological lesions and the selective sensitivity of (sometimes small) regions. In practice a good scheme, whether designed for toxicology,

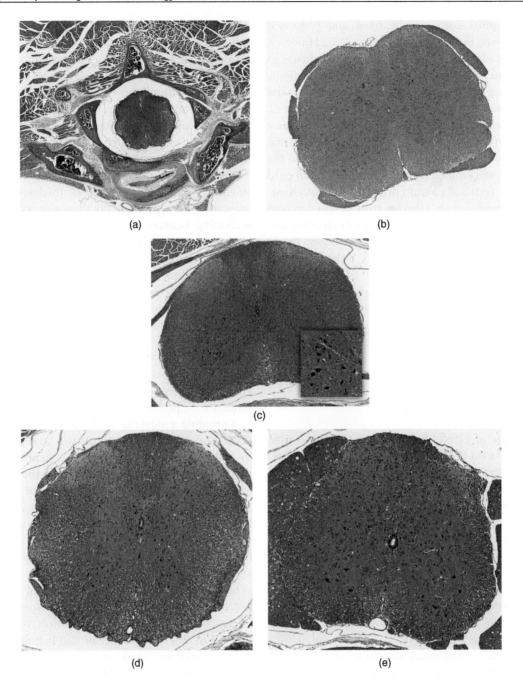

Figure 8.5   Spinal cord fixation, artefacts and histology. The spinal cord can be fixed and processed *in situ* (a, (c), (d) and (e)) with the vertebral column (after immersion in decalcification fixative) or fixed in NBF and isolated from the vertebral column prior to processing (b). The former may lead to artefactual changes. These include retraction of the cord in the vertebral canal (a), which can also be appreciated from the wrinkled surface of the cord and dark neurons (visible in (a), (c), (d), (e) and in insert in (C)). The illustrations in (c), (d) and (e) also provide a comparison of the relative size and shape of the central grey matter and dorsal horns in the cervical (c), thoracic (d) and lumbar (e) spinal cord. In cross section, the cervical cord is flatted dorsoventrally, the thoracic cord is more circular and the lumbar cord more trapezoid with proportionally thinner white matter and increasingly large surrounding nerve bundles as one progresses caudally. (Note: the scale varies in these different images)

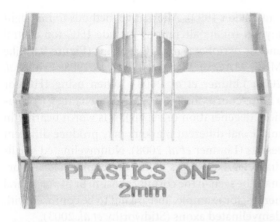

Figure 8.6 Brain matrix to aid reproducible cutting of brain slices.

phenotyping or fundamental research will cover a selection of anatomical/functional regions, be easily reproducible and consistent across the number of animals in the study and may also depend on the study focus.

Most trimming schemes used by pathologists consist of coronal sections which are easily compared with atlas plates (Morawietz *et al.* 2004; Jordan *et al.* 2011; Switzer *et al.* 2011; Bolon *et al.* 2013) (Figure 8.3 and Table 8.1). Sagittal/ parasagittal sections can be used to examine specific structures but can be difficult to obtain with consistent orientations because of their length.

Once the brain has been sliced, it is good practice to lay the different samples produced on the working surface prior to embedding the tissue in their cassettes and ensure that all are present and that the cranio-caudal orientation of the embedded surface is known. Microscopic sections are only few microns thick and as most trimming schemes will be limited to less than ten sections, flipping one of the trimmed fragment will result in examining the same section twice instead of two different sections a few millimetres apart. When trying to correlate clinical signs with any potential unilateral histological lesion, the preservation of the left/right orientation is also essential and would also be lost if the segments are inadvertently flipped at the time of embedding.

In the absence of visible macroscopic change or when carrying a routine sampling scheme for microscopic examination of the spinal cord, one to three segments of cord (from the cervical, thoracic and lumbar region) are considered sufficient. Transverse sections of cord provide a cross-section of nerves fascicules while longitudinal sections allow nerves to be followed and facilitate the identification of lesions. On a transverse section, the shape of the central grey matter makes it possible to recognize cervical, thoracic and lumbar areas (Figure 8.5) (Radovsky and Mahler 1999). A comprehensive scheme can include both transverse and longitudinal sections. The cervical intumescences, brachial plexus and nerves extending from the cord in the sacral region are not routinely examined. Labelling the tissue to preserve the craniocaudal and therefore left/right orientation is important to locate the changes (cervical, thoracic, lumbar, left or right, ventral or dorsal) and, if relevant, associate them with the relevant clinical signs. This can be done by having a small incision, a suture or an ink spot placed on one side, which would be recognizable on the histological section.

Dorsal root ganglia can be readily examined when the spinal cord has been fixed in situ in the vertebral canal (Figure 8.5). Serial parasagittal or horizontal sections can also be prepared to maximize the chance for DRGs to be included. Alternatively they can be sampled from the cord at necropsy and embedded in a cassette in groups of representative cervical, thoracic and lumbar ganglia (Bolon *et al.* 2013).

8.5   Special stains and techniques

The landmarks visible on the trimmed sections and additional ones can readily be distinguished on H&E-stained sections. Brain atlases often show Nissl's stain (Cresyl violet) sections (instead of H&E sections), which distinguishes white and grey matter clearly (Figure 8.7). Nissl's stain or toluidine blue stained sections can be used when more detailed structures needs to be distinguished, as discrete grey matter nuclei and white matter tracts can be

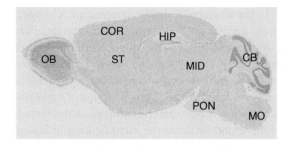

Figure 8.7 Parasagittal section of Nissl's stained mouse brain. Nissl's stain (Cresyl violet) detects neurons by staining purple-blue the Nissl's substance (RER) they contain in their cytoplasm. This therefore provides a good contrast between highly stained grey matter and paler, more poorly stained white matter. On this section the olfactory bulbs (OB), cerebral cortex (COR), striatum (ST), hippocampus (HIP), midbrain (MID), pons (PON), cerebellum (CB) and medulla oblongata (MO) regions can readily be identified.

more readily identified and, from their respective locations, precise position can be inferred with the help of an atlas. However the examination of H&E sections provides sufficient detail for identifying a number of structures and to locate the occurrence of potential pathological changes. Also it offers better cell morphology for the characterization of these lesions. Routine H&E sections should be 3–5 um thick as thicker sections sometimes used in research settings can give the tissue a hypercellular appearance, which can be confusing on histological examination (Figure 8.8).

Additional histochemistry and IHC stains can also complement H&E routine examination. Similar to Nissl or toluidine blue staining, Luxol fast blue (LFB) distinguishes myelin with good contrast by staining it blue (Figure 8.8) and is often used to infer the presence of demyelinated lesions from a loss of staining in the white matter. However, alternative pathological changes (like inflammatory cell infiltration and/or oedema) can also result in a loss of LFB staining. For this reason, when myelin is the focus of the examination, EM remains the gold standard technique. Black gold (Figure 8.8), Oil red O and Sudan black stain lipid-rich myelin, but because lipids are dissolved during the solvent-based processing of paraffin embedded tissues, these stains cannot be used on paraffin block samples (Schmued

and Slikker 1999). Alternative methods to highlight myelin (or its alteration) include IHC for central and/or peripheral myelin proteins (Figure 8.8) like MBP, MAG or P0 (Martini *et al.* 1988; Erb *et al.* 2006; Lindner *et al.* 2008). When using IHC for a time-course analysis of myelin changes (formation, degeneration or repair), it is worth bearing in mind that different markers may produce different results (Lindner *et al.* 2008). Nonmyelinated small-diameter axons (less than 1 um diameter) can normally be seen in the corpus callosum or in peripheral nerves, for example, and are not to be confused with demyelinated axons (Stidworthy *et al.* 2003).

Degenerate neurons can be picked up on an H&E-stained section, where they appear hypereosinophilic with dark shrunken nuclei (they are sometimes referred to as "red dead" neurons) (Figure 8.9). Additional histochemistry stains, which include Amino cupric silver (ACS) or fluorojade (FJ), are considered to help confirm an H&E diagnosis and provide an increased contrast and sensitivity to discriminate between affected and nonaffected neurons (for example where degenerate neurons are sparse). Amino cupric silver is challenging technically and when a fluorescent microscope is available (with a FITC filter fitted), the FJ staining protocol is easier to apply and can be combined with IF markers for double or triple labelling (Krinke *et al.* 2001; Garman 2011, Jordan *et al.* 2011; Quin and Crews 2012). Fluorojade (either B or C) works on FFPE tissue following immersion fixation but it is important to note that red blood cells (in immersion fixed or poorly perfused tissues) will also fluoresce green with FJ (Bolon *et al.* 2008; Jordan *et al.* 2011). It is also important to include positive control material when using these special stains although it can be difficult to find good examples.

When a particular subpopulation of interest is to be examined (i.e. neurons, astocytes, oligodendrocytes, macrophages), specific additional stains can be used. Common cell type specific IHC markers include neuron-specific enolase (NSE) for the recognition of neurons; glial fibrillary acidic protein (GFAP) for astrocytes (Figure 8.10) (with astrocyte activation manifesting by an increase in cell size, number of processes and intensity

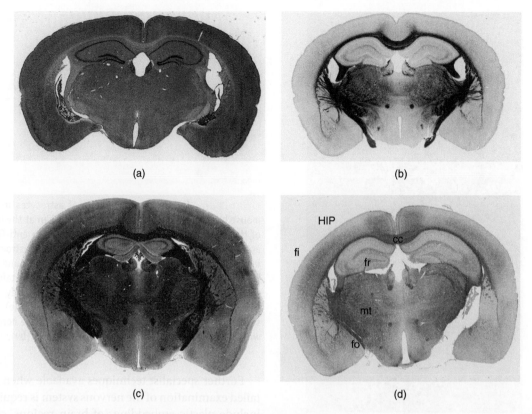

Figure 8.8 Section thickness and additional myelin stains. Routine H&E sections are generally 3–5 um thick. Thicker sections, like here a 30 um section (a) can give the tissue a hypercellular appearance (compare with routine 3–5 um H&E-stained section (b) from Figure 8.11). White matter can be distinguished from grey matter regions on an H&E section (a) thereby providing a number of recognizable landmarks for reference. Other myelin stains can be used to complement and H&E section when a more detailed examination of the myelin is required. These include black gold (b), Luxol fast blue (c) or myelin basic protein (MBP) immunohistochemistry (d). White matter tracts which are visible in this brain section at the hippocampus level (HIP) include the corpus callosum (cc), the fasciculus retroflexus (fr), the mammillothalamic tract (mt), the fornix (fo) and the fimbria (fi). These appear bright eosinophilic on H&E, bright read on a black gold, dark blue on a Luxol fast blue, and brown on this MBP IHC.

of the GFAP staining), CD68/ED1 or ionized calcium binding adaptor molecule (Iba1) for microglia/macrophages. Markers for distinguishing between microglia and macrophage subpopulations have also been described (Kofler and Wiley 2011). Myelinating cells: oligodendrocytes (found in the CNS enwrapping multiple axons) and Schwann cells (found in the PNS enwrapping individual axons) can be labelled by IHC using myelin protein markers. Other markers that label the various stage of glial cell progenitors are described in the literature, although they are rarely used as part

of the common histopathology tool kit. Specific subpopulations of neurons can also be mapped using IHC for distinct neurotransmitters. Tyrosine hydroxylase (TH), for instance, stains dopaminergic neurons and is readily applied on FFPE tissues (Zeiss 2005). In the field of neuroscience, IHC is often carried out using immunofluorescent markers (IF), which allows double or triple labelling of the cell population(s) of interest. For pathological examination, however, the disadvantage is that cell morphology is not visible by IF. At best, a DAPI or Hoechst labelling of nuclei gives some indication of

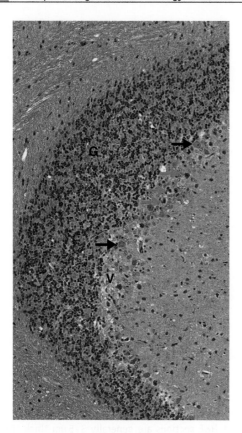

Figure 8.9 Neuronal necrosis in the brain. Necrotic neurons (red neurons) can be picked up on an H&E stained section, where they appear hypereosinophilic with dark shrunken nuclei (compare with the appearance of dark neuron artefact on Figures 8.5 and 8.16). Here in this cerebellar section necrotic Purkinje neurons (blue arrows) are amid morphologically unremarkable ones (black arrows). This is accompanied by minimal vacuolation of the neuropil (V). The granular layer (G) appears unaffected.

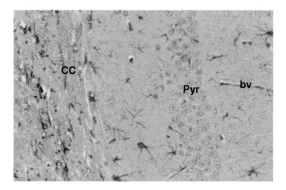

Figure 8.10 Immunohistochemistry of astrocytes in the mouse brain. This section of mouse brain, taken at the level of the hippocampus and immunostained with an anti-GFAP antibody, shows brown-stained, stellar-shaped astrocytes in the white matter of the corpus callosum (cc) and the pyramidal neuron layer (Pyr) of the hippocampus containing GFAP protein in the cytoplasm of the main body and processes of cells (cell nuclei are counterstained blue). The stained processes of the cells can be seen in close contact with pyramidal neurons and along blood vessels (bv).

the cell density. Although less common than dual or triple IF labelling, multiplexing is also achievable using non-fluorescent chromogens (van der Loos 2008). IHC with chromogens/DAB often provides as good a labelling and can be counterstained for examination using standard light microscopy, making morphological assessment of the tissue and of the antigen distribution easier. Although perfusion fixed frozen sections are often used in neuroscience protocols, most markers will also work on FFPE tissues.

Further specialist techniques available when detailed examination of the nervous system is required include plastic embedding (of brain regions, cord or nerves) and teased nerve preparation (Krinke *et al.* 2000b). Plastic embedding is mostly used as a preliminary step to EM examination and contrary to paraffin embedding processing does not dissolve the lipid-rich myelin. Classically, semi-thin (1 um) sections are cut, stained with toluidine blue and examined by light microscopy to select the region of interest to be examined by EM. These semi-thin sections often offer exquisite details for light microscopy examination and can be valuable in their own right.

Among other techniques, quantification, which is not specific to the nervous system, is commonly used in neurosciences and the reader is referred to Chapter 2 for more information on this topic. Quantifying cell number and volume is often considered as it can provide a surrogate measure of degeneration and in particular neurodegeneration. Alterations in neurotransmitter levels which can be of functional consequence are not necessarily accompanied by overt morphological changes on H&E examination

(Zeiss 2005). Also following a wave of cell death there may be no resulting morphological alterations except for a decreased number of neurons and subtle gliosis. In these cases quantification methods, can be required to complement H&E examination.

8.6   Microscopic examination

Familiarity with the common landmarks seen on macroscopic or microscopic sections helps to avoid confusing them with lesions. It can also be helpful to develop a checklist or a map of structures as a neuroanatomy guide to the sections examined and an aide-memoire of the areas which have been looked at (section 8.2, 8.4, Table 8.1 and Figure 8.11). In addition to the white matter tracts and grey matter neuronal cell bodies, it is important to be familiar with the other cell types within the CNS. The ventricular system with its bilaterally symmetrical lateral ventricles merging into the 3^d (dorsal and ventral) and 4th ventricles caudally is lined by ciliated ependymal cells and richly vascular choroid plexi (Figure 8.12). Just beneath the ependymal layers of the lateral ventricles are progenitor cell rich clusters of the subventricular zone (Figure 8.12) (Hagan *et al.* 2011; Pastrana *et al.* 2011; Fuentealba *et al.* 2012; Walton 2012). Other cell rests are also sometimes present: for example on the surface of the cerebellum (Hagan *et al.* 2011) (Figure 8.12). The circumventricular organs (Figure 8.12) are another set of structures, which, given their small size, may not be consistently present on H&E sections from all animals examined. In close relation with the ventricular system, they include secretory and sensory organs: the vascular organ of the lamina terminalis, the subfornical organ, the median eminence, the subcommissural organ, the area postrema and the pineal gland (Radovsky and Mahler 1999; Wohlsein *et al.* 2013). The neurohypophysis and choroid plexi are also sometimes included in this list of circumventricular organs. Another anatomical structure which should not to be confused with a pathological change is the sharp transition between

the oligodendrocyte myelinated axons of the CNS and Schwann cell myelinated axons of the PNS whether in the cranial or in the peripheral nerves (Garman 2011).

8.7   Normal histology-juvenile – CNS and PNS

Specific morphological features of the embryo and juvenile nervous system are worthy of mention. Waves of neurogenesis, synaptogenesis, apoptosis, loss of synapses (pruning), gliogenesis and glial cell differentiation take place during development, and these also are not to be confused with pathological changes. In the mouse they continue during postnatal development, where in addition to intrinsic regulation (that started in the embryonic phase), external sensory stimulations post partum also plays a role (Kato *et al.* 2012). Post natal growth and maturation of the nervous system in juvenile mice during the first weeks of life is reflected morphologically by an increase in the brain size and weight (Ye *et al.* 2002; Fu *et al.* 2013) and functionally by the progressive evolution of sensory-motor capacities in the young animals.

From an embryological perspective, the CNS derives from the neural grove of the ectoderm (neural ectoderm or neuroepithelium), which forms around embryonic day 7. As the embryo completes its 'turning', the neural grove that runs from the future head to the tail region progressively closes caudally and cranially from day embryonic day 8 to 9 to form the neural tube (Kaufman 1992). Adjacent neural crest regions differentiate into the future cranial nerves and DRGs and are generally well differentiated by embryonic day 11–11.5. Sympathetic trunks/ganglia only appear from day 10.5 and peripheral nerves are not readily distinguishable until day 15 (Kaufman 1992).

The cranial region of the neural tube forms 3 primitive brain vesicles (fore-, mid- and hindbrain). Regions of the brain are often referred to according to this embryological development and further subdivisions from these three primitive vesicles are

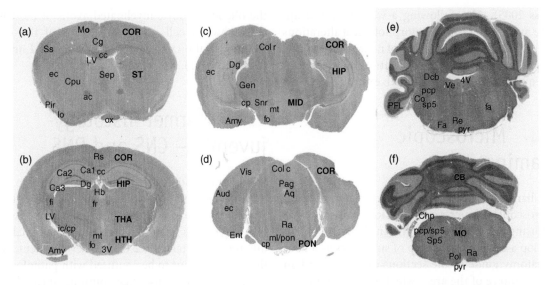

Figure 8.11 Six-section trimming pattern of the mouse brain and microscopic landmark identification. On an H&E coronal sections set numerous anatomical and functional structures can be identified. The areas of the cerebral cortex (COR) with their concentric cortical nuclear layers can be subdivided into functional regions: dorsally the frontal and parietal cortex including motor (Mo) and somatosensory (Ss) areas, along the dorsal median fissure the cingulate (Cg) and retrosplenial (Rs) cortex, ventrally the piriform cortex (Pir) and caudally the visual region (Vis) of the occipital cortex and the auditory (Aud), amygdala (Amy) and entorhinal regions (Ent) of the temporal cortex. Subcortical structures include the striatum (ST) composed mainly of the caudate nucleus and putamen (Cpu) separated by a white matter tract, the internal capsule (ic). Several other distinct white matter tracts can easily be recognized on H&E sections including the anterior commissure (ac), a bilateral tract fusing caudally; the lateral bilateral olfactory tracts (lo), the corpus callosum (cc), external capsule (ec) and fimbria (fi) which respectively run transversally above, laterally from and cranially of the hippocampus. Further caudally the bilateral fasciculus retroflexus (fr), the mammillothalamic tracts (mt), the more ventral fornix (fo), the ventrolateral cerebral peduncles (cp), the transversal medial lemniscus (ml) and transversal fibres of the pons (pon) can be found. Other internal microscopic landmarks include the hippocampus (HIP) with its Ammon horns (Ca1, Ca2 and Ca3) and dentate gyrus (Dg), the habenular nucleus (Hb), the thalamus (THA) and, below it, the hypothalamus (HTH) sitting above the optic chiasm (ox). The geniculate nuclei (Gen) sit laterally of the midbrain (MID), which contains the substantia nigra (Snr), the rostral (Col r) and caudal (Col c) colliculi (i.e. superior and inferior). The cerebellum (CB) is foliated, each folia containing distinct molecular, granular and Purkinje cell layers. At its level several distinct mostly bilateral cerebellar nuclei can be differentiated: the deep cerebellar (Dcb), cochlear (Co), vestibular (Ve), facial (Fa), reticular (Re) and trigeminal (Sp5) nuclei, the raphe (Ra) and posterior olive (Pol). These are interspersed among large identifiable white matter tracts: the inferior (caudal or posterior) cerebellar peduncles (pcp), spinal tract of the trigeminal nucleus (sp5), pyramids (pyr) and facial tracts (fa). Other recognisable structures on H&E sections include the choroid plexi (Chp) and the ventricular system of the lateral (LV), 3rd (3V) and 4th (4V) ventricles separated by the aqueduct of sylvius (Aq) itself surrounded by periaqueductal grey matter (Pag).

recognized. From the forebrain (proencephalon) evolve 2 lateral telencephalons (from which the cerebral hemispheres will evolve) and a central diencephalon (lining the third ventricle), while the hindbrain (rhombencephalon) is separated into a rostral metencephalon (i.e. the future pons and cerebellum) and a caudal myelencephalon (i.e. the medulla oblongata). Also evolving from the forebrain vesicle are the optic vesicles, the third ventricle, the lamina terminalis and the infundibulum (from which the pituitary develops) (Figure 8.13).

From day 11–11.5, three concentric layers of the neural tube can be distinguished: the marginal layer (which in the spinal cord will give rise to white matter regions), the mantle layer (the future grey matter in the spinal cord) and, closest to the central canal,

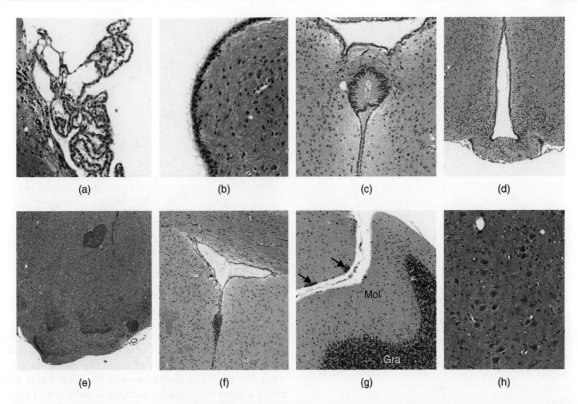

Figure 8.12 Diversity of the cell types identified histologically in the mouse brain. The choroid plexi producing the CSF can be found in the ventricles as well-vascularized connective tissue lined by a cuboidal epithelium (a). The ventricles are lined by cuboidal and ciliated ependymal cells (b). The distinct appearance of the circumventricular organs is exemplified in the subcommissural organ (c) and the median eminence (d). Other anatomical structures on H&E sections not to be confused with lesions include clusters of granular neurons in the basal forebrain at the level of the olfactory tubercle (islands of Calleja) (e) and remaining precursor of the subventricular zone (f). Other cell remnants can also sometimes be seen on the surface of the cerebellar folia forming small clusters at the periphery of the molecular layer of the cerebellum (arrows in g). The molecular (Mol), Purkinje cell (Pur) and granular cell layers (Gra) that form the cerebellar folia can also be seen here. The diversity in the appearance of neurons can also be appreciated by comparing neurons in the molecular, Purkinje and granular cell layers (g) with the large polygonal neurons with prominent nucleoli found, for example, in the deep cerebellar nucleus (h) (Note: the scale varies in these different images).

the ependymal (ventricular zone) layer (Figure 8.13). The differentiation into alar plate (dorsally) and basal plate (ventrally) giving rise, respectively, to the dorsal and ventral horns in the spinal cord, becomes visible later, around day 13 (Kaufman 1992). Overall, the spatial position of cells in the neural tube determines their differentiation fate. Dorsoventral and anteroposterior patterning are under the influence of several intrinsic signalling cascades, like sonic hedgehog, bone morphogenetic proteins, Wnt and retinoic acid. These have been well characterized in the spinal cord where a bilateral groove in the neural tube (the sulcus limitans) separates a dorsal sensory and a ventral motor area. Patterning also regulates cell lineage differentiation. Oligodendrocytes for instance, originate from several spatially and temporally distinct ventral and dorsal pools of proliferative, migratory precursors starting from around day 13 (Kessaris *et al.* 2008; Fancy *et al.* 2011).

In the brain, important structures start to differentiate from about day 12: thalamus, hypothalamus, striatum, choroid plexi, olfactory lobes and nerves.

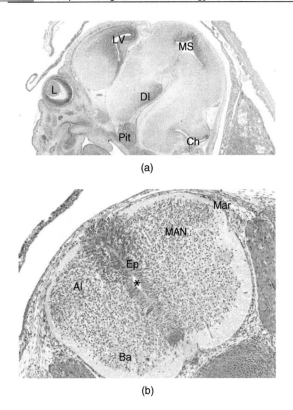

(a)

(b)

Figure 8.13 Brain and spinal cord regions of a mouse embryo at day 14.5. (a) The lateral ventricle (LV) lined by the highly cellular future cerebral cortex, the mesencephalic vesicle (MS) covered by the highly cellular midbrain roof, the choroid plexus protruding in the 4th ventricle, the edge of the diencephalon (DI), the residual lumen of the Rathke's pouch in the region of the future pituitary (Pit) and the eye with the lens (L) at its centre can be identified in the head region of the embryo. (b) A cross section through the neural tube at the level of the spinal cord in a same age embryo reveals the central canal (*) surrounded by the concentric layers of the ependymal (Ep), the mantle (MAN) and the marginal (Mar) layers. The dorsal alar (Al) and ventral basal (Ba) plates are also indicated.

Glioblasts and neuroblasts differentiate and migrate as the cortical layer forms around day 14, shortly followed by differentiation of the cerebellum on day 15. The pineal gland does not appear from the diencephalon until day 14.5. By embryonic day 17, most of the brain, spinal cord (including distinguishable grey and white matter and cervical, thoracic and lumbar regions), peripheral nerves and

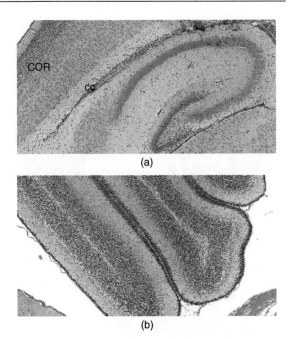

(a)

(b)

Figure 8.14 Hippocampus of a juvenile mouse on post natal day 4 and cerebellum of a juvenile mouse on post natal day 12. (a) In comparison with the adult mouse the tissue appears more cellular, particularly white matter areas like the corpus callosum (cc) lying above the hippocampus. The cortical neuronal layers (COR) also appear condensed. (b) Likewise the cerebellar folia (here on day 12) are highly cellular, with little myelin, a relatively thin molecular layer when compared with the granular layer and an additional superficial granular layer of which remnants can be seen in the adult (compare with Figure 8.11).

autonomic system are in place in the embryo and continue to differentiate through day 18 and after birth (Kaufman 1992).

On an H&E section, when compared with that of an adult animal, the brain and spinal cord of a late embryo or juvenile mouse will appear hypercellular (Figure 8.14), with some variations in the relative proportions of the different regions (white matter areas are proportionally smaller as myelination does not occur until the first weeks post partum). Despite the expression of myelin protein genes in the embryo and shortly after birth, little myelin is detected in PND 7 mice brains (Verity and Campagnoni 1988; Ye *et al.* 2002; Bansal 2003; Oh *et al.* 2003), although it can be detected earlier in the spinal cord

(Erb 2006). In the brain, myelination peaks in the 3d week of life and is considered complete by week 10 (Verity and Campagnoni 1988; Ye *et al.* 2002). Axon maturation and myelination involve active communication processes between neurons and glia (Emery 2010; Doretto *et al.* 2011). This is exemplified in the proportional relationship between the thickness of the myelin sheath formed by the myelinating cells and the size of the axons they enwrap (Fancy *et al.* 2011). Also taking place during the first postnatal weeks are further waves of apoptosis (Naruse and Keiro 1995), synaptogenesis (Paolicelli *et al.* 2011) and neurogenesis (Fu *et al.* 2013; Kato *et al.* 2012). These can be appreciated in the evolution of the respective sizes, proportions and positions of neuronal layers in the hippocampus and in the cerebellar folia (where an external granular layer is present during the first weeks of life) (Figure 8.14). Neurogenesis even carries on during adulthood. The process is well described in the olfactory bulbs where new neurons mature from progenitors located in the subventricular zone and migrate along the rostral migratory stream (Pastrana *et al.* 2011; Fuentealba *et al.* 2012; Walton 2012).

8.8   Artefacts – CNS and PNS

Many artefacts are not CNS-specific and are discussed elsewhere in this book and in the literature (McInnes 2005). Artefacts of importance in the nervous tissue are dark neurons and vacuolation because these can interfere with pathological examination and represent a diagnostic challenge.

At necropsy, there are two main ways to induce artefacts. One is by inadvertently sectioning the brain (generally the cerebral cortex or sometimes the cerebellum) and/or inserting bone fragments into it (Figure 8.15) when opening the skull. The other is by crushing and or pressing the tissue (with a finger, an instrument or a bone fragment). At its worst, crushing results in loss of the tissue architecture, but pressure applied on unfixed tissue can also cause microscopic artefacts (dark neurons) (Figure 8.16).

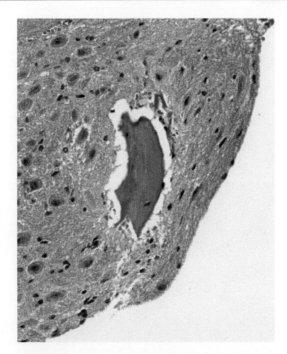

Figure 8.15   Artefactual introduction of a bone fragment in the brain.

Small, dark basophilic neurons with intensely stained shrunken nuclei seen in brain or spinal cord sections are often referred to as dark neurons (Jortner 2006; Garman 2011; Kaufmann *et al.* 2012; Wohlsein *et al.* 2013). Although they can affect any neuronal population, they are most commonly seen in large neurons, particularly in the hippocampus, the cerebral cortex or the cerebellum, where affected cells often cluster (Figure 8.16). The causes discussed above can in part be alleviated by perfusion fixation; however, dark neurons do occasionally also occur in perfusion fixed tissue, especially if the perfusion results in incomplete tissue fixation (Wohlsein *et al.* 2013). The diagnostic challenge is to differentiate these neurons from necrotic ones, which are also shrunken but in addition show a condensed nucleus and a bright eosinophilic cytoplasm (see also section 8.5 and Figure 8.9). This can be achieved by getting familiar with the respective appearances of dark and red neurons, while being aware that some of the literature can be confusing with dark neurons having been reported as neurodegeneration.

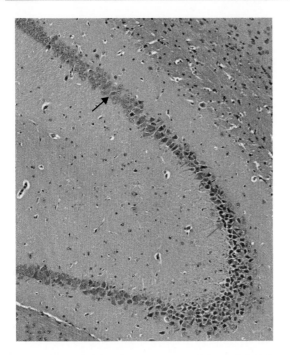

Figure 8.16   Dark neuron artefact. A cluster of dark basophilic neurons (blue arrow) in the pyramidal neurons of the hippocampus is common. The dark shrunken basophilic (stained dark blue) neurons with prominent corkscrew-shaped dendrites are adjacent to other intact neurons (black arrow) (compare with necrotic neurons in Figure 8.9).

Complementary techniques (IHC, fluorojade staining or ACS) described in section 8.5 can be used to differentiate the two. Finally the presence of concurrent controls prepared and processed simultaneously is in practice often relied upon to help distinguish artefact from pathological changes.

Vacuolation is another diagnostic dilemma in neuropathology that can be associated both with artefact and pathological changes (Figures 8.4 and 8.17) (Radovsky and Mahler 1999). It is often subdivided on the basis of its location (intracellular or extracellular, neuronal, myelinic, astrocytic) and on the size of the vacuoles (macro or microvesicular vacuolation). Although this can be of assistance to differentiate artefact from lesion, each case should be considered carefully. The lack of associated morphological changes (cellular debris and/or associated gliosis) or region specific distribution can be suggestive of artefact, although this is not always true, as acute or locally extensive vacuolation can occur in some lesions (Wozniak *et al.* 1996). In these cases, an awareness of the clinical signs and or the duration of the experiment regimen and the presence of concurrent controls, prepared and processed simultaneously and consistently remain of invaluable diagnostic assistance. Artefactual vacuolation can result from autolysis, improper tissue handling at necropsy, prolonged immersion in alcohol baths during processing (for example, over the weekend) (Kaufman *et al.* 2012) or on frozen fixed tissue (McInnes *et al.* 2005) and techniques to minimize its occurrence have been described (Fix and Garman 2000). Vacuolation, which is sometimes reported as a pathological change in DRGs or autonomic ganglia, can also be artefactual (Rogers-Cotrone *et al.* 2010; Butt 2010; Jortner and Rogers-Cotrone 2011). (Figure 8.17).

Other less commonly seen artefacts include pale white matter staining areas resulting from insufficient dehydration (McInnes 2005) (Figure 8.18) or the use of cold perfusion fixatives (Fix and Garman 2000) and the rarely described pale blue amorphous PAS positive bodies called mucocytes or Buscaino bodies, which are considered secondary to an unusual reaction between myelin and fixatives (Fix and Garman 2000; McInnes 2005; Wohlsein *et al.* 2013).

8.9   Pathological changes and background pathology – CNS and PNS

The terminology used to describe morphological changes associated with pathological processes in the CNS can vary from that used in other organs or for the normal nervous tissue. For example, enlarged activated astrocytes with eosinophilic cytoplasm are referred to as gemistocytes, foamy macrophages as Gitter cells, and elongated microglia as rod cells. Chromatolysis describes the dispersion of the Nissl's substance (the RER, which normally gives a granular appearance to the cytoplasm of neurons) and satellitosis describes a group of glial cells encircling

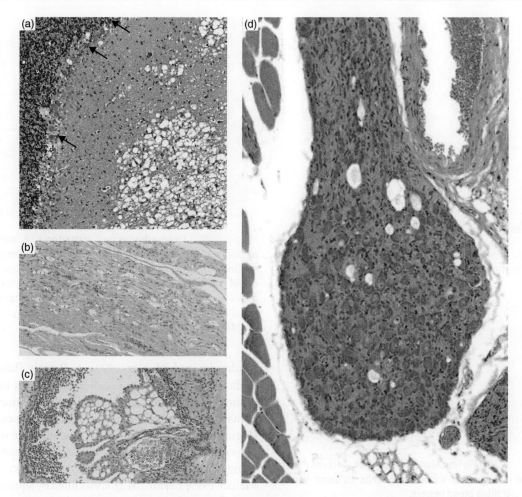

Figure 8.17 Vacuolation in the CNS. (a) This large area of vacuolation in the cerebellum is accompanied by more subtle vacuolation in the Purkinje cell layer where necrotic neurons can also be seen (arrows). This is considered pathological. Compare with the artefactual vacuolation presented in Figure 8.4, where no associated pathological change was seen. (b) Vacuolation in the spinal cord is readily identified on a longitudinal section where lines of vacuoles can be seen which represents degeneration of individual myelinated axons (Wallerian degeneration). This is accompanied by the presence of cellular debris and occasional inflammatory cells in what is sometimes called digestion chambers. This is considered pathological. (c) Vacuolation of the choroid plexus associated with inflammatory cell infiltrate in the ventricle (left) and in the neuropil (right) and also considered pathological. (d) Vacuolation can occasionally be seen in the DRGs of normal animals in the absence of any pathological changes.

a neuron. Both changes are considered to indicate neuronal degeneration, although they can also occur as a normal finding (Wohlsein *et al.* 2013). Perivascular cuffing refers to inflammatory cell infiltration around a blood vessel (in the Virchow-Robin space) (Figure 8.19).

A number of pathological changes are commonly seen as background in mice. The nature and

incidence of these changes can vary across strains, sexes and laboratories and it is worth maintaining a database of incidences in wild type or control animals.

Common congenital changes include hypocallosity and hydrocephalus. Hypocallosity or aplasia of the corpus callosum is common (up to 70%) in some strains including BALB/c and 129 (Percy

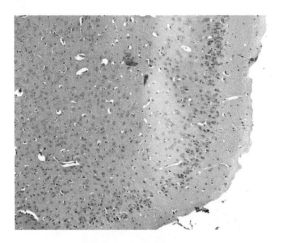

Figure 8.18   Artefactual pale white matter staining areas resulting from insufficient dehydration.

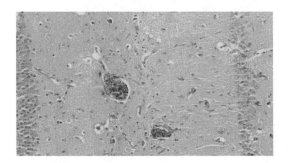

Figure 8.19   Perivascular cuffing. Two foci of tightly packed mononuclear inflammatory cells surround two red blood cell filled blood vessels.

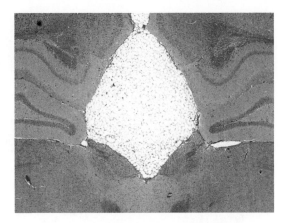

Figure 8.20   Lipoma/lipogenous hamartoma. A mass of adipocytes expands from the ventricular space along the midline here slightly compressing the adjacent hippocampus of both sides, Adipocytes give the mass a honeycombed appearance as the lipid content of the cytoplasm has been dissolved during processing leaving only the polygonal to rounded cell membranes visible.

and Barthold 2007) and congenital hydrocephalus (ventricular enlargement) is common among C57BL strains (Radovsky and Mahler 1999; Vogel *et al.* 2012). Affected animals may not display clinical signs until weaning (when they are unable to independently eat and drink) or may even be clinically silent (Percy and Barthold 2007). In genetically engineered mice (GEMs), high levels of Cre recombinase expression in neuronal precursors have also been linked with hydrocephalus, microencephaly and secondary defects in ependymal lining (Percy and Barthold 2007). Ventricular dilation (hydrocephalus) can also be seen in older animals secondary to compression by a pituitary mass leading to CSF accumulation

(Radovsky and Mahler 1999). Cell rests (heterotopias) (Figure 8.12) are occasionally present in the brain (hippocampus and cerebellum). Lipomatous hamartoma (sometimes referred to as lipomas) (Figure 8.20), which consist of unremarkable white adipose tissue often sit in between hemispheres along the midline or ventricles (Radovsky and Mahler 1999; Taylor 2012). Cysts lined by squamous epithelium infrequently occur in the brain or cord (Radovsky and Mahler 1999; Taylor 2012). Extramedullary haematopoiesis, which is common in mice in various tissues, can also occasionally be seen around choroid plexi. Meningeal melanosis can be found in pigmented mice including C57BL/6 (Percy and Barthold 2007) (Figure 8.21). Spontaneous lipofuscin accumulation in neurons or glia is rare in rodents (Kaufmann *et al.* 2012).

Spontaneous infectious diseases resulting in inflammatory lesions (Figure 8.22) are rare in modern laboratory animal facilities and most infectious agents resulting in CNS lesions do so as a result of experimental rather than natural infection (i.e. encephalitis with intracerebral injection of norovirus, mouse hepatitis virus or Theiler's virus) (Radovsky and Mahler 1999). It is always worth being aware

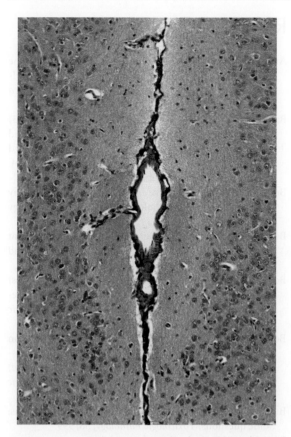

Figure 8.21   Meningeal melanosis. Meningeal cells cytoplasm is heavily filled with melanin granules giving the meninges surrounding the central patent blood vessel its dark brown colour.

that unusual lesions due to infection can occur in immunocompromised mice.

When carrying out a pathological examination of tissues from an animal model (transgenic or induced) it is important to be aware of the reported changes previously observed in the background of this strain, in the model itself and whether any of these are affected by the age or the sex of the animals (Sundberg *et al.* 1991; Vogel *et al.* 2012). Hydrocephalus, for example, which can be congenital can also be seen following exposure to curpizone (a toxic mouse model of demyelination) at doses marginally higher to those used to induce demyelination (Kesterson and Carlton 1970). Likewise, spontaneous or induced seizures can occur in

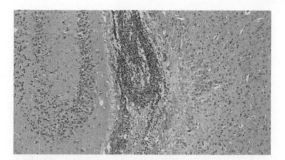

Figure 8.22   Brain inflammation. A heavy infiltrate of inflammatory cells and cell debris adjacent to the hippocampus.

mice with variable strain and age susceptibility (for example, young DBA and FVB and aging BALBc) and may or may not be associated with morphological changes (Goelz *et al.* 1998; McLin and Steward 2006; Percy and Barthold 2007). Where associated pathology is seen, its extent can be variable, consisting of neuronal necrosis with or without gliosis, located in the cortex, hippocampus and thalamus.

In aging mice, mineralized foci, often located in the mid brain, are common (Radovsky and Mahler 1999; Percy and Barthold 2007; Taylor 2012). These are laminated and have smooth edges (Figure 8.23)

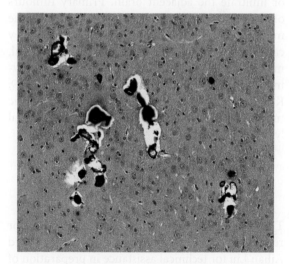

Figure 8.23   Thalamic mineralization. Mineralized foci are densely stained, acellular and randomly distributed in the neuropil of the thalamus. The neuropil retraction around these foci is artefactual.

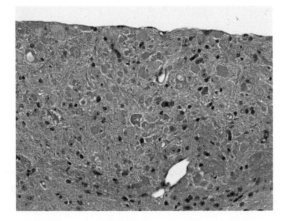

Figure 8.24 Axonal dystrophy. Cross-section of axons in the dorsal part of the brain stem. They are variably sized and shaped, hypereosinophilic (bright pink-red) and some appear slightly vacuolated.

as opposed to the artefactual sharp bone fragments that can be introduced at necropsy (Figure 8.15). Another common aging change is the neuroaxonal dystrophy observed in some deep nuclei of the CNS, often located in the brain stem (Figure 8.24) (Radovsky and Mahler 1999). Pituitary tumours, which are common in female mice, can compress or infiltrate the adjacent brain. Primary tumours of the nervous system, however, are rare in the most commonly used mouse strains, although they can develop from any cell lineage (neural, glial, mixed, undifferentiated, choroid plexus, ependymal or meningeal) (Radovsky and Mahler 1999; Krinke *et al.* 2000b). Multicentric and/or metastatic tumours also occur on occasion (e.g. meningeal lymphoma infiltration).

8.10   Acknowledgment

The author gratefully acknowledges Brad Bolon for review of the manuscript and John Bowles and Ailian Liu for technical assistance in preparation of images for this chapter.

References

Auer, R.N. and Coulter, K.C. (1994) The nature and time course of neuronal vacuolation induced by the N-methyl-D-aspartate antagonist MK-801. *Acta Neuropathologica* **87**, 1.

Bansal, R., Lakhina, V., Remedios, R. and Tole, S. (2003) Expression of FGF receptors 1, 2, 3 in the embryonic and postnatal mouse brain compared with Pdgfrα, Olig2 and Plp/dm20: implications for oligodendrocyte development. *Developmental Neuroscience* **25**, 83.

Bolon, B. (2000) Comparative and correlative neuroanatomy for the toxicologic pathologist. *Toxicologic Pathology* **28**, 6.

Bolon, B., Anthony, D.C., Butt, M. *et al.* (2008) 'Current pathology techniques.' Symposium review: advances and issues in neuropathology. *Toxicologic Pathology* **36**, 871.

Bolon, B., Garman, R., Jensen, K. *et al.* (2006) A 'best practices' approach to neuropathologic assessment in developmental neurotoxicity testing for today. *Toxicologic Pathology* **34**, 296.

Bolon, B., Garman, R.H., Pardo, I.D. *et al.* (2013) STP position paper: recommended practices for sampling and processing the nervous system (brain, spinal cord, nerve, and eye) during nonclinical general toxicity studies. *Toxicologic Pathology* **41**, 1028.

Butt, M.T. (2010) Vacuoles in dorsal root ganglia neurons: Some questions. *Toxicologic Pathology*, **38**, 999.

Butt, M. (2011) Evaluation of the adult nervous system in preclinical studies. In *Fundamental Neuropathology for Pathologists and Toxicologists: Principles and Techniques*, Wiley-Blackwell, Hoboken, NJ, p. 323.

Doretto, S., Malerba, M., Ramos, M. *et al.* (2011) Oligodendrocytes as regulators of neuronal networks during early postnatal development. *PLoS ONE* **6**, e19849.

Emery, B. (2010). Regulation of oligodendrocyte differentiation and myelination. *Science* **330**, 779.

Erb, M., Flueck, B., Kern, F. *et al.* (2006) Unraveling the differential expression of the two isoforms of myelin-associated glycoprotein in a mouse expressing GFP-tagged S-MAG specifically regulated and targeted into the different myelin compartments. *Molecular and Cellular Neuroscience* **31**, 613.

Fancy, S.P., Chan, J.R., Baranzini, S.E. *et al.* (2011) Myelin regeneration: a recapitulation of development? *Annual Review of Neuroscience* **34**, 21.

Fix, A.S. and Garman, R.H. (2000) Practical aspects of neuropathology: a technical guide for working with the nervous system. *Toxicologic Pathology* **28**, 122.

Fu, Y., Rusznák, Z., Herculano-Houzel, S. *et al.* (2013) Cellular composition characterizing postnatal development and maturation of the mouse brain and spinal cord. *Brain Structure and Function* **218**, 1337.

Fuentealba, L.C., Obernier, K. and Alvarez-Buylla, A. (2012) Adult neural stem cells bridge their Niche. *Cell Stem Cell* **10**, 698.

Garman, R.H. (2011) Histology of the central nervous system. *Toxicologic Pathology* **39**, 22.

Goelz, M.F., Mahler, J., Harry, J. *et al.* (1998) Neuropathologic findings associated with seizures in FVB mice. *Laboratory Animal Science* **48**, 34.

Hagan, C.E., Bolon, B., Keene, C.D. (2011) Nervous system, in *Comparative Anatomy and Histology: A Mouse and Human Atlas* (eds P.M. Treuting, S. Dintzis, D. Liggitt and C.W. Frevert), Academic Press, London, p. 339.

Hale, S.L., Andrew-Jones, L., Jordan, W.H. *et al.* (2011) Modern pathology methods for neural investigations. *Toxicologic Pathology* **39**, 52.

Irvine, K. and Blakemore, W.F. (2008) Remyelination protects axons from demyelination-associated axon degeneration. *Brain* **131**, 1464.

Jordan, W.H., Young, J.K., Hyten, M.J. and Hall, D.G. (2011) Preparation and analysis of the central nervous system. *Toxicologic Pathology* **39**, 58.

Jortner, B.S. (2006) The return of the dark neuron. A histological artefact complicating contemporary neurotoxicologic evaluation. *Neurotoxicology* **27**, 628. Comment in *Neurotoxicology* **27**, 1126.

Jortner, B S (2011) Preparation and analysis of the peripheral nervous system. *Toxicologic Pathology* **39**, 66.

Jortner, B, Rogers-Cotrone, T. (2011) Vacuoles in dorsal root Ganglia neurons: some questions. *Toxicol Pathol.* **39**, 451.

Kato, Y., Kaneko, N., Sawada, M. *et al.* (2012) A subtype-specific critical period for neurogenesis in the postnatal development of mouse olfactory glomeruli. *PLOS One* **7**, 1.

Kaufman, M.H. (1992) *The Atlas of Mouse Development*, 1st edn, Academic Press, London.

Kaufmann, W., Bolon, B., Bradley, A. *et al.* (2012) Proliferative and nonproliferative lesions of the rat and mouse central and peripheral nervous systems. *Toxicologic Pathology* **40**, 87S.

Kessaris, N., Pringle, N., Richardson, W.D. (2008) Specification of CNS glia from neural stem cells in the embryonic neuroepithelium. *Philosopical Transactions of the Royal Society of London. Series B, Biological sciences* **363**, 71.

Kesterson, J.W., Carlton, W.W. (1970) Aqueductal stenosis as the cause of hydrocephalus in mice fed the substituted hydrazine, cuprizone. *Experimental and Molecular Pathology* **13**, 281.

Knoblaugh, S., Randolph-Habecker, J., Rath, S. (2011) Necropsy and histology, in Comparative Anatomy and Histology: A Mouse and Human Atlas (ed. P.M. Treuting, S. Dintzis, D. Liggitt and C.W. Frevert), Academic Press, London, p. 15.

Kofler, J. and Wiley, C.A. (2011) Microglia: key innate immune cells of the brain. *Toxicologic Pathology* **39**, 103.

Krinke, G.J., Classen, W., Vidotto, N. *et al.* (2001) Detecting necrotic neurons with fluoro-jade stain. *Experimental Toxicologic Pathology* **53**, 365.

Krinke, G.J., Kaufmann, W., Mahrous, A.T. and Schaetti, P. (2000a) Morphologic characterization of spontaneous nervous system tumors in mice and rats. *Toxicologic Pathology* **28**, 178.

Krinke, G.J., Vidotto, N. and Weber, E. (2000b) Teased-fiber technique for peripheral myelinated nerves: methodology and interpretation. *Toxicologic Pathology* **28**, 113.

Lindner, M., Heine, S., Haastert, K. *et al.* (2008) Sequential myelin protein expression during remyelination reveals fast and efficient repair after central nervous system demyelination. *Neuropathology and Applied Neurobiology* **34**, 105.

Martini, R., Bollensen, E. and Schachner, M. (1988) Immunocytological localization of the major peripheral nervous system glycoprotein P0 and the L2/HNK-1 and L3 carbohydrate structures in developing and adult mouse sciatic nerve. *Developmental Biology* **129**, 330.

McInnes, E. (2005) Artefacts in histopathology. *Comparative Clinical Pathology* **13**, 100.

McLin, J.P. and Steward, O. (2006) Comparison of seizure phenotype and neurodegeneration induced by systemic kainic acid in inbred, outbred, and hybrid mouse strains. *European Journal of Neuroscience* **24**, 2191.

Morawietz, G., Ruehl-Fehlert, C., Kittel, B. *et al.* (2004) Revised guides for organ sampling and trimming in rats and mice – Part 3. *Experimental Toxicologic Pathology* **55**, 433.

Naruse, I. and Keiro, H. (1995) Apoptosis in the developing CNS. *Progress in Neurobiology* **47**, 135.

Oh, L.Y., Denninger, A., Colvin, J.S. *et al.* (2003) Fibroblast growth factor receptor 3 signaling regulates the onset of oligodendrocyte terminal differentiation. *Journal of Neuroscience* **23**, 883.

Paolicelli, R.C., Bolasco, G., Pagani, F. *et al.* (2011) Synaptic pruning by microglia is necessary for normal brain development. *Science* **333**, 1456.

Pastrana, E.E.E., Silva-Vargas, V. and Doetsch, F. (2011) Eyes wide open: a critical review of sphere-formation as an assay for stem cells. *Cell Stem Cell* **8**, 486.

Paxinos, G. and Franklin, K.B.J. (2012) *Paxinos and Franklin's the Mouse Brain in Stereotaxic Coordinates*, 4th edn, Academic Press, London.

Percy, D.H. and Barthold, S.W. (2007) *Pathology of Laboratory Rodents and Rabbits*, 3rd edn, Wiley-Blackwell, Hoboken, NJ.

Quin, L. and Crews, F.T. (2012) Chronic ethanol increases systemic TLR3 agonist-induced neuroinflammation and neurodegeneration. *Journal of Neuroinflammation* **9**, 130.

Radovsky, A. and Mahler, J.F. (1999) Nervous system, in *Pathology of the Mouse: Reference and Atlas* (ed. R. Maronpot), Cache River Press, Vienna.

Rao, D.B., Little, P., Malarkey, D.E. *et al.* (2011) Histopathological evaluation of the nervous system in national toxicology program rodent studies: a modified approach. *Toxicologic Pathology* **39**, 463.

Rogers-Cotrone T., Burgess M.P., Hancock S.H., Hinckley J., Lowe K., Ehrich M.F., Jortner B.S. (2010) Vacuolation of sensory ganglion neuron cytoplasm in rats with long-term exposure to organophosphates. *Toxicologic Pathology*, **38**, 554.

Schmued, L. and Slikker, W. (1999) Black-gold: a simple, high-resolution histochemical label for normal and pathological myelin in brain tissue sections. *Brain Research* **837**, 289.

Shimeld, C., Efstathiou, S. and Hill, T. (2001) Tracking the spread of a lacZ-tagged herpes simplex virus type 1 between the eye and the nervous system of the mouse: comparison of primary and recurrent infection. *Journal of Virology* **75**, 5252.

Spijker, S. (2011) Dissection of rodent brain regions. *Neuroproteomics Neuromethods* **57**, 13.

Sundberg, J.P., Woolcott, B.L., Cunlifee-Beamer, T. *et al.* (1991) Spontaneous hydrocephalus in inbred strains of mice. *JAX Notes*, http://jaxmice.jax.org/jaxnotes/archive/445a.html (accessed 19 July 2013).

Stidworthy, M., Genoud, S., Suter, U. *et al.* (2003) Quantifying the early stages of remyelination following cuprizone-induced demyelination. *Brain Pathology* **13**, 329.

Switzer, R.C., Lowry-Franseen, C. and Benkovic, S.A. (2011) Recommended neuroanatomical sampling practices for comprehensive brain evaluation in nonclinical safety studies. *Toxicologic Pathology* **39**, 73.

Taylor, I. (2012) Mouse, in *Background Lesions in Laboratory Animals. A Color Atlas* (ed. E.F. McInnes), Saunders Elsevier, London.

van der Loos, C.M. (2008) Multiple immunoenzyme staining: methods and visualizations for the observation with spectral imaging. *Journal of Histochemistry and Cytochemistry* **56**, 313.

Verity, A.N. and Campagnoni, A.T. (1988) Regional expression of myelin protein genes in the developing mouse brain: in situ hybridization studies. *Journal of Neuroscience Research* **21**, 238.

Vogel, P., Read, R.W., Hansen, G.M. *et al.* (2012) Congenital hydrocephalus in genetically engineered mice. *Veterinary Pathology* **49**, 166.

Walton, R.M. (2012) Postnatal neurogenesis: of mice, men, and macaques. *Veterinary Pathology* **49**, 155.

Wohlsein, P., Deschl, U. and Baumgärtner, W. (2013) Non-lesions, unusual cell types, and postmortem artifacts in the central nervous system of domestic animals. *Veterinary Pathology* **50**, 122.

Wozniak, D.F, .Brosnan-Watters, G., Nardi, A. *et al.* (1996) MK-801 neurotoxicity in male mice: histologic effects and chronic impairment in spatial learning. *Brain Research* **707**, 165.

Ye, P., Li, L., Richards, R.G., DiAugustine, R.P. and D'Ercole, A.J. (2002) Myelination is altered in insulin-like growth factor-I null mutant mice. *Journal of Neuroscience* **22**, 6041.

Zeiss, C.J. (2005) Neuroanatomical phenotyping in the mouse: the dopaminergic system. *Veterinary Pathology* **42**, 753.

Useful website resources

Brain histology, http://ctrgenpath.net/static/atlas/mousehistology/Windows/nervous/diagrams.html (accessed 19 July 2013).

Brain atlas, http://www.hms.harvard.edu/research/brain/atlas.html (accessed 19 July 2013).

Brain maps, http://brainmaps.org/index.php?p=speciesdata&species=mus-musculus (accessed 19 July 2013).

The mouse brain library, http://www.mbl.org (accessed 19 July 2013).

The e-mouse atlas project, http://www.emouseatlas.org/emap/home.html (accessed 19 July 2013).

The mouse genome informatics, http://www.informatics.jax.org/ (accessed 19 July 2013).

Spinal cord, http://www.christopherreeve.org/site/c.ddJFKRNoFiG/b.4427053/k.F400/Spinal_Cord_Atlas_Introduction_and_Mouse_Spinal_Cord.htm#Mousecervicalsections (accessed 19 July 2013); http://reni.item.fraunhofer.de/reni/trimming/index.php (accessed 19 July 2013).

Chapter 9
Lymphoid and haematopoietic system

Ian Taylor

Huntingdon Life Sciences, Eye, Suffolk, UK

9.1 Introduction

The lymphoid and haematopoietic system is composed of multiple organs and tissues distributed throughout the body and is responsible for the development of the immune response, and the production of the blood's cellular components.

The primary lymphoid organs are the bone marrow and thymus, responsible for production and maturation of the B- and T- lymphocytes respectively. The secondary lymphoid organs, the lymph nodes, spleen and mucosa-associated lymphoid tissues maintain populations of mature lymphocytes and are the sites of antigenic stimulation and clonal expansion. The lymphoid organs are composed of two tissue components, reticular connective tissue and lymphatic tissue, composed of lymphocytes, macrophages and antigen presenting cells.

Haematopoietic stem cells are active in the mouse liver from embryonic day 10, and in the spleen from embryonic day 13, but the bone marrow becomes the primary site of haematopoiesis from embryonic day 18 onwards (Holsapple *et al.* 2003). The predominant site of haematopoiesis in the adult mouse is the bone marrow of the long bones, but the spleen retains haematopoietic activity throughout adult life and haematopoietic activity in the liver can be seen in response to disease (Haley 2003; Taylor 2011; Linden *et al.* 2012).

The anatomy and histology of the structural components of each of the different lymphoid tissues are presented below but there are several sources of information (some of which are available online) that provide details of the basic anatomy of the mouse (Cook 1965, 1983; Hummel *et al.* 1966), and provide guidance on general necropsy and histology practices, which include descriptions of procedures to follow for the dissection and trimming of these tissues. Reviews of the normal histopathology of the lymphoid system of the mouse, and changes

A Practical Guide to the Histology of the Mouse, First Edition. Cheryl L. Scudamore.
© 2014 John Wiley & Sons, Ltd. Illustrations © Veterinary Path Illustrations, unless stated otherwise.
Published 2014 by John Wiley & Sons, Ltd. Companion Website: www.wiley.com/go/scudamore/mousehistology

associated with exposure to xenobiotics in preclinical toxicity tests are also available (Ward *et al.* 1999; Maronpot 2006).

There are complex interactions between the different organs of the lymphoid system, which is a dynamic system reacting to changes in antigenic stimulation throughout life. These reactions can be manifested as morphological changes in the different components of the system. There can also be pronounced strain-, genetic-, age-, and sex-dependent variations in the function (Sellers *et al.* 2012) and normal appearance of lymphoid organs, which need to be taken into account when performing histopathological evaluation of these tissues (Elmore 2012). Careful comparison of findings seen in lymphoid tissues with the range of changes observed in concurrent control animals is required to ensure that normal background variations are taken into account when interpreting these findings (Figure 9.1). It can also be difficult to differentiate

between severe hyperplastic reactions in lymphoid tissues and lymphoma or leukaemia. Knowledge of the range of reactive changes in the species being examined, and of species-, sex- and strain-related differences in anatomical and histological appearance of the lymphoid tissues, as well as an understanding of the biological behaviour of the neoplastic changes being diagnosed is important in reaching correct diagnoses (Ward *et al.* 2012).

9.2   Lymph nodes

9.2.1   *Background and development*

Lymph nodes develop from embryonic mesenchyme forming a bud, which is enveloped by a lymphatic sac at around embryonic day 16 and 17, remaining as primitive tissue until after birth (Ward *et al.* 1999; Willard-Mack 2006). Stem cells from the thymus and bone marrow populate the fibrovascular tissue of the node, and development of germinal centres and proliferative activity continues after birth. The distinction between cortex and medulla is not visible until day 4 postnatally, and complete morphological development occurs around a month after birth (van Rees *et al.* 1996).

9.2.2   *Sampling techniques*

Because of the small size of murine lymph nodes, care has to be taken to ensure consistency of trimming and orientation. The majority of mouse lymph nodes may be sampled and sectioned complete (Seymour *et al.* 2004), and Morawietz *et al.* (2004) recommend longitudinal sections of the whole organ to allow examination of all major areas. Consistent trimming of larger lymph nodes should be employed to avoid introducing artefactual variations in structure (Elmore 2012). Formalin fixation is generally acceptable for most routine morphological investigations of lymphoid organs however alternative fixation techniques may need to be considered if

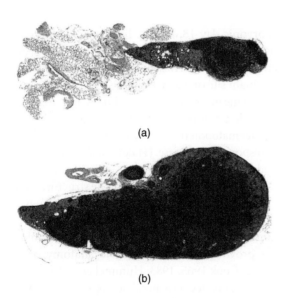

(a)

(b)

Figure 9.1   Comparison of morphology of axillary lymph nodes from control mice at 20 weeks of age, illustrating the type of variation that can be seen in the size of the nodes and the proportions of the different components: (a) relatively small axillary lymph node with few primary follicles; (b) relatively large axillary lymph node with prominent paracortical areas, secondary follicles, sinus erythrocytosis and pigmented macrophages.

immunohistochemistry needs to be used (Mikaelien *et al.* 2004). Immunohistochemistry may be particularly useful for investigating and confirming shifts in populations of immune cell subsets in lymphoid tissues (Ward *et al.* 2006, Rehg *et al.* 2012) and diagnosing neoplasia. Having basic immunohistochemistry protocols for labelling T and B cells and macrophages in formalin fixed paraffin embedded tissue (Rehg *et al.* 2012) available in the laboratory may be useful for early investigation of a phenotypic change but more specific protocols for different immature and mature immune subsets will need to be worked up on a case by case basis using knowledge of the experimental protocol and any genetic modifications.

9.2.3 Artefacts

As mentioned above, inappropriate trimming or sectioning of small lymph nodes can introduce artefactual variation in apparent size of the different compartments of the lymph node (Figure 9.2). The presence of erythrocytes within the sinusoids can be indicative of haemorrhage in the tissues drained by the lymph node, but this change can

Figure 9.3   The presence of blood within the sinuses of an axillary lymph node associated with agonal haemorrhage or damage during necropsy procedures.

also be agonal or artefactual as a result of damage during necropsy procedures (Figure 9.3). The presence of haemosiderin laden macrophages and erythrophagocytosis is indicative of the ante-mortem nature of the change (Elmore 2006a; Taylor 2011). Crush artefacts of lymph nodes are a result of inappropriate handling at necropsy or histology (Figure 9.4).

9.2.4 Anatomy and histology

Correct identification and nomenclature of the murine lymph nodes is important to ensure

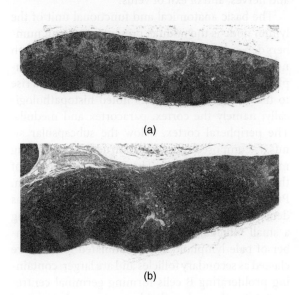

(a)

(b)

Figure 9.2   Variation in morphology of lymph node associated with off-centre (a) or central (b) sectioning of the node.

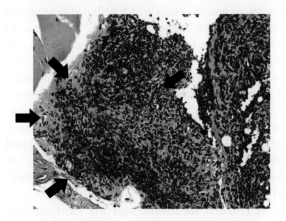

Figure 9.4   Crush artefact of lymph node due to inappropriate handling of fresh or fixed tissue. Smearing of the lymphocytes in affected region (arrows).

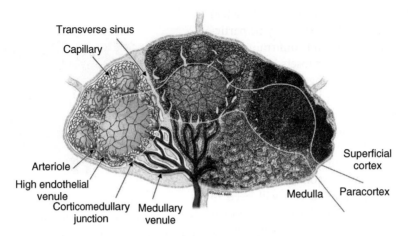

Transverse sinus

Capillary

Arteriole

High endothelial
venule

Corticomedullary
junction

Medullary
venule

Superficial
cortex

Medulla   Paracortex

Figure 9.5   Representation of the structure of a lymph node, illustrating on the left the vascular structure, in the centre, the reticular meshwork superimposed on the vasculature, and on the right, a lobule as it appears histopathologically. (Reprinted by permission of SAGE Publications from Willard-Mack, C.L. (2006) Normal structure, function, and histology of lymph nodes. *Toxicologic Pathology*, **34**, 409–424, and by permission of the illustrator, David A Sabio.)

reproducibility of findings in scientific studies (Van den Broeck *et al.* 2006) and a clear understanding of lymph node structure, and regional variations in structure, is important in correctly identifying changes in the nodes (Willard-Mack 2006). A generalized illustration of the structure of a 'typical' lymph node is presented in Figure 9.5.

There are relatively few lymph nodes in the mouse compared to other species (Haley 2003). Van den Broeck *et al.* (2006) identified 22 different nodes in the mouse. These are bean-shaped structures connected to lymph vessels and distributed throughout the body. They are small and can be difficult to identify within fat and connective tissues. The peripheral lymph nodes are bilateral and include the mandibular, axillary and popliteal lymph nodes. Central lymph nodes in the thoracic or abdominal cavities are usually not bilateral (including the mediastinal, pancreatic and mesenteric) (Van den Broeck *et al.* 2006; Linden *et al.* 2012). The distribution and nomenclature of the lymph nodes in the mouse has been well described by Van den Broeck *et al.* (2006) (Figure 1.9 and Table 9.1).

A reticular meshwork supports the lymphatic components and the lymph node is composed of lymphoid lobules, separated by lymph filled sinuses and surrounded by a fibrous capsule (Linden

et al. 2012). Afferent lymph vessels enter the lymph node at the convex edge, connecting to subcapsular sinuses and draining through the cortex and paracortex into the medullary sinuses, and draining from the lymph node via efferent lymphatics at the hilus. The hilus is also the point of entry of arterioles and nerves, and of exit of veins.

The basic anatomical and functional unit of the lymph node is the nodule, present in varying numbers dependent on the size and location of the lymph node. The differentiation of structures and lymphoid cell populations within the nodules gives rise to the structural differences noted histopathologically, namely the cortex, paracortex and medulla. The peripheral cortex, below the subcapsular sinus, is composed of follicular structures consisting mainly of B lymphocytes. The size and appearance of the follicles is dependent on antigenic stimulation. Primary follicles are unstimulated and appear as dense collections of small lymphocytes surrounding a small follicular centre containing a small number of pale lymphoid cells. Stimulated follicles are classed as secondary follicles and are larger, containing proliferating B cells forming germinal centres containing large lymphoblasts, macrophages and tingible body macrophages (Figure 9.6). The paracortex represents the area of the lymph nodes

Table 9.1   Lymph nodes.

English name	Official name	Occurrence (F/E)*	Topography
1. Mandibular lymph node (ln.)	Ln. mandibularis	Constant (21/21)	Rostromedial to the sublingual and mandibular salivary glands
2. Accessory mandibular ln.	Ln. mandibularis accessorius	Constant (21/21)	Dorsolateral to the mandibular lymph node
3. Superficial parotid ln.	Ln. parotideus superficialis	Constant (21/21)	Ventral to the external acoustic pore, caudal to the extraorbital lacrimal gland, cranioventral to the parotid salivary gland, dorsal to the junction between the superficial temporal vein (v.) and the maxillary.
4. Cranial deep cervical ln.	Ln. cervicalis profundus cranialis	Constant (24/24)	Medial to the external jugular vein and sternocephalic muscle (m.), lateral to sternohyoid m., caudal to digastric m., dorsal to the trachea
5. Proper axillary ln.	Ln. axillaris proprius	Constant (12/12)	Medial to the shoulder, dorsolateral to ascending pectoral m., at the junction between the lateral thoracic vein and the axillary vein
6. Accessory axillary ln.	Ln. axillaris accessorius	Constant (12/12)	Caudal to triceps brachii m., lateral to cutaneous trunci m., in subcutaneous adipose tissue
7. Subiliac ln.	Ln. subiliacus	Constant (12/12)	In the fold of the flank (plica lateralis) cranial to thigh musculature, near the deep circumflex iliac artery (a.) and v.
8. Sciatic ln.	Ln. ischiadicus	Constant (12/12)	Medial to superficial gluteus m., caudal to gluteus medius m. and sciatic nerve
9. Popliteal ln.	Ln. popliteus	Constant (12/12)	In the popliteal fossa between biceps femoris m. and semitendinosus m.
10. Cranial mediastinal lnn.	Lnn. mediastinales craniales	Constant (3/3)	Bilaterally two lymph nodes located lateral to the thoracic thymus and along the internal thoracic a. and v.
11. Tracheobronchal ln.	Ln. tracheobronchalis	Constant (6/6)	Single (unpaired) lymph node at the tracheal bifurcation
12. Caudal mediastinal ln.	Ln. mediastinalis caudalis	Constant (3/3)	Single (unpaired) lymph node in the caudal mediastinum, ventral to the esophagus, along the ventral vagal trunk
13. Gastric ln.	Ln. gastricus	Constant (24/24)	Single (unpaired) lymph node in the lesser omentum at the minor curvature of the stomach
14. Pancreaticoduodenal ln.	Ln. pancreaticoduo-denalis	Constant (24/24)	Single (unpaired) lymph node in the mesoduodenum, dorsal to the portal vein, surrounded by pancreatic tissue
15. Jejunal lnn.	Lnn. jejunales	Constant (24/24)	Large cluster of lymph nodes in the mesojejunum along the cranial mesenteric a.
16. Colic ln.	Ln. colicus	Constant (24/24)	In the mesocolon at the transition between ascending colon and transverse colon
17. Caudal mesenteric ln.	Ln. mesentericus caudalis	Constant (24/24)	Single (unpaired) lymph node in the caudal mesentery at the origin of the caudal mesenteric a.
18. Renal ln.	Ln. renalis	Constant (33/33)	Dorsal to the ipsilateral kidney nearby the renal blood vessels, caudal to the adrenal gland

(continued overleaf)

Table 9.1   (*continued*)

English name	Official name	Occurrence (F/E)*	Topography
19. Lumbar aortic ln.	Ln. lumbalis aorticus	Inconstant (4/6 bilateral, 2/6 only left)	Lateral to (and adjacent with) the abdominal aorta, halfway between the origin of the renal and common iliac arteries
20. Lateral iliac ln.	Ln. iliacus lateralis	Inconstant (2/3 only right, 1/3 absent)	In adipose tissue caudolateral to the kidney along the deep circumflex iliac a.
21. Medial iliac ln.	Ln. iliacus medialis	Constant (21/21)	Major bilateral lymph node at the terminal segment of the abdominal aorta and the origin of the common iliac a.
22. External iliac ln.	Ln. iliacus externus	Constant (3/3)	Small lymph node along the initial (intra-abdominal) segment of the external iliac a., before the latter enters the femoral canal

*F: number of animals in which lymph nodes were found; E: number of animals in which these particular lymph nodes were examined.
(Reprinted with permission from Elsevier from Van den Broeck, W., Derore, A. and Simoens, P. (2006) Anatomy and nomenclature of murine lymph nodes: Descriptive study and nomenclatory standardization in BALB/cAnNCrl mice. *J Immunol Methods*, **312**, 12–19.)

containing predominantly T lymphocytes, and is situated between the follicles and the medullary sinuses. Antigenic stimulation of T lymphocytes leads to proliferation of these cells in the paracortex but without the formation of follicular structures or germinal centres. High endothelial venules (HEV) are the site of entry of vascular lymphocytes into the stroma of the lymph nodes (Figure 9.7). They are located throughout the interfollicular cortex and paracortex but appear more obvious at the periphery of the paracortex. As these vessels transition into the medulla, the high endothelium is lost and they become lined by squamous endothelium typical of the medullary venules. The medulla is composed of cords and sinuses, with variable numbers of lymphocytes, plasma cells and macrophages. Plasma cell precursors from the cortex migrate to the medulla following B-cell stimulation, where they mature and they release antibodies into the lymph. Following antigenic stimulation, the cords can be packed with plasma cells and small lymphocytes (Figure 9.8). A more detailed description of the anatomy of the lymph node and the nodular structure in relation to immune function is presented in the review by Willard-Mack (2006).

A variety of spontaneous changes are seen in the lymph nodes of mice, some of which represent changes related to antigenic stimulation, often of an unknown aetiology. Proliferative changes may be secondary to the presence of tumours or degenerative changes in the animal, particularly the presence of skin lesions which can result in hyperplasia of cellular elements in draining lymph nodes. The incidence of lymphocyte hyperplasia (Figure 9.9) increases with age and tends to be more common in females than males (Frith *et al.* 2001; Elmore 2006a). Plasma cell hyperplasia and mast cell hyperplasia (Figure 9.10) are also commonly seen but there are strain differences in the background incidence of mast cells in lymph nodes (Ward *et al.* 1999; Frith *et al.* 2001; Elmore 2006a).

Atrophy is often seen in the lymph nodes of ageing mice (Ward *et al.* 1999) and can be associated with fatty infiltration of the lymph node (Figure 9.11), or sinus ectasia (Elmore 2006a; Taylor 2011) (Figure 9.12). Angiectasis is most often seen in the mesenteric lymph nodes and can be distinguished from sinus ectasia by the presence of endothelial lining cells (Figure 9.13). Sinus erythrocytosis can be related to the lymph node draining a site of haemorrhage but can also be artefactual. The levels of erythrophagocytosis and haemosiderin pigment seen are indicative of the chronicity of the lesion (Elmore 2006a; Taylor 2011) (Figure 9.14). Sinus histiocytosis is characterized by the presence of large macrophages with eosinophilic cytoplasm

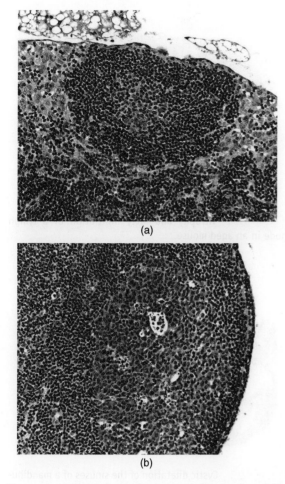

(a)

(b)

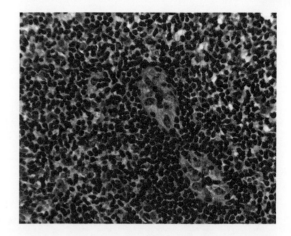

Figure 9.7   High endothelial venules.

Figure 9.6   Comparison of primary and secondary follicles in lymph nodes. (a) Collection of small lymphocytes surrounding a small follicular centre containing a small number of pale lymphoid cells; (b) Large secondary follicle with increased number of pale staining cells and tingible body macrophages.

Figure 9.8   Plasmacytosis. Medullary sinuses packed with plasma cells in an enlarged mandibular lymph node.

in the medullary and subcapsular sinuses of the lymph nodes, often associated with phagocytosis of pigment (commonly haemosiderin or lipofuschin) or other materials (Figure 9.15). Hyperplasia of dendritic reticular cells is occasionally seen in young mice and is considered to be secondary to viral infections (Taylor 2011) (Figure 9.16). Extramedullary haematopoiesis may occasionally be seen in lymph nodes, secondary to haemorrhage or severe inflammation (Frith *et al.* 2001; Elmore 2006a; Taylor 2011). The lymph nodes are described

as a common site for deposition of amyloid in different strains of mice, particularly the CD-1 mouse (Elmore 2006a; Percy and Barthold 2007). However, the occurrence of amyloidosis in CD-1 mice has shown time-related changes and can vary considerably between animal suppliers (Maita *et al.* 1988; Engelhardt *et al.* 1993; Taylor 2011).

Neoplastic changes of the lymph nodes, other than secondary involvement of haematopoietic neoplasms are rarely seen in mice and almost never in mice less than a year old. Haemangiomas are occasionally seen in lymph nodes in chronic toxicity studies (Faccini *et al.* 1990).

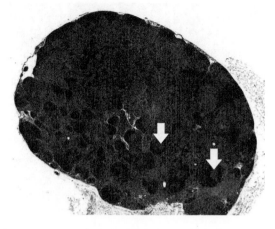

Figure 9.9   Lymphocytic hyperplasia of the lymph node with an overall increase in lymphocytic-rich compartments and increased germinal centres (arrows).

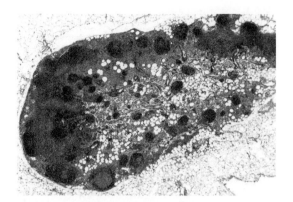

Figure 9.11   Atrophy and fatty infiltration of a lymph node in an aged mouse.

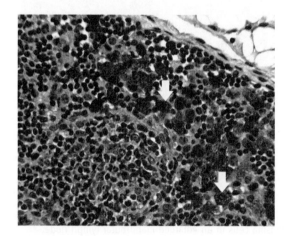

Figure 9.10   Increased heavily granulated mast cells (arrows) in the medullary sinuses of a lymph node.

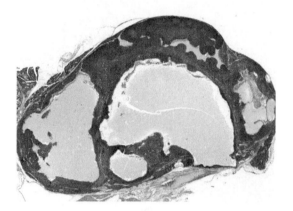

Figure 9.12   Cystic dilatation of the sinuses of a mandibular lymph node.

9.3   Spleen

9.3.1   *Background and development*

The spleen is involved in the filtration of blood to remove effete red blood cells, and is the second largest lymphoid organ, mediating immune responses to blood borne antigens. The red pulp of the spleen is also a normal site of extramedullary haematopoiesis

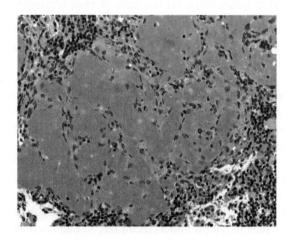

Figure 9.13   Angiectasis in the mesenteric lymph node of an aged CD-1 mouse.

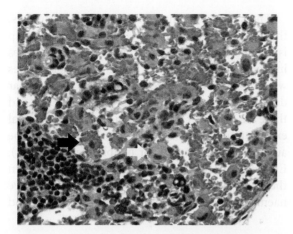

Figure 9.14   Sinus erythrocytosis with erythrophagocytosis (black arrow) of the axillary lymph node. Presence of haemosiderin pigment in macrophages (white arrow).

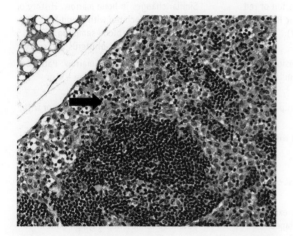

Figure 9.15   Sinus histiocytosis of a mesenteric lymph node. The sinuses contain increased numbers of large, pale eosinophilic histiocytes (arrow).

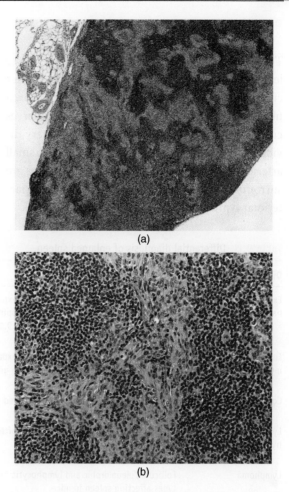

(a)

(b)

Figure 9.16   Hyperplasia of dendritic reticular cells. (a) Low power view of axillary lymph node showing increased numbers of eosinophilic cells within sinusoids. (b) High power view illustrating eosinophilic cells with elongated nuclei.

9.3.2   Sampling techniques

Sampling the spleen transversely at the widest part of the organ tends to give the most repeatable sections, allowing evaluation of all the structural components of the spleen (Morawietz *et al.* 2004; Elmore 2012). Some authors recommend longitudinal sections to ensure that a sufficient amount of white pulp is available for evaluation (Suttie 2006) but it is generally harder to achieve consistent sections to allow for semiquantitative comparisons between mice using this technique.

in the mouse (van Rees *et al.* 1996; Suttie 2006; Linden *et al.* 2012).

The development of the spleen can be identified on embryonic day 13, originating from mesenchyme in the dorsal mesogastrium, which forms the reticular structure of the spleen. The spleen is populated by stem cells from the bone marrow and thymus by embryonic day 15 (van Rees *et al.* 1996). By day 17 the spleen is grossly similar to the adult spleen, but active lymphoid tissue does not develop until after birth (Ward *et al.* 1999).

9.3.3 *Anatomy and histology*

The spleen is an elongated organ, lying in the upper left quadrant of the abdomen, alongside the greater curvature of the stomach. It is surrounded by a fibrous capsule, and the splenic artery and nerves enter the spleen at the hilus. The splenic artery divides into trabecular arteries and branches into small arterioles throughout the spleen. The arterioles are surrounded by lymphoid tissue, forming the periarteriolar lymphoid sheaths (PALS) (Cesta 2006a).

The spleen is composed of two compartments, the red pulp and the white pulp. The red pulp is composed of a vascular network and sinuses involved in filtration of the blood, and macrophages which phagocytose foreign material and cellular debris. The spleen is often enlarged at necropsy and there are a number of physiological and pathological explanations for this so this change should not be automatically mistaken for lymphoma (Table 9.2). Extramedullary haematopoiesis is common in the red pulp of the mouse spleen and reactive haematopoiesis involving myeloid, erythroid and

Table 9.2 Differential diagnosis of enlarged spleen.

Diagnosis	Morphological appearance	Commonly related features in other organs
Extramedullary haematopoiesis	Regular enlargement of spleen due to expansion of red pulp by increased number of small basophilic RBC precursors and megakaryocytes (Figure 9.17).	Similar changes in bone marrow. History or pathology associated with blood loss – e.g. trauma, recurrent blood sampling, haemorrhagic lesions of uterus and ovary.
Increased myelopoiesis	Regular enlargement of spleen due to expansion of white pulp by increased numbers of immature granulocyte forms and band cells.	Similar features in bone marrow. Pathology associated with inflammation – e.g. neoplasia, abscess, ulceration.
Congestion	Regular enlargement of spleen due to red pulp expanded by mature RBCs.	Sometimes seen as agonal change or with barbiturate anaesthesia
Lymphoid hyperplasia	Expansion of white pulp components, often with increased germinal follicle formation.	Similar changes in other lymphoid organs. Chronic inflammation or neoplasia in other organs.
Lymphoma	Follicular/pleomorphic and lymphocytic most common types affecting spleen in mice.	
	Follicular/pleomorphic lymphoma – massive expansion of white pulp by pseudonodules of irregular shaped lymphocytes and macrophages (Figure 9.18).	Similar lesions may be seen in Peyer's patches and mesenteric lymph nodes.
	Lymphocytic lymphoma – initially diffuse expansion of white pulp by regular dark staining lymphocytes can eventually affect whole organ. Loss of normal architecture.	Commonly also involve thymus, liver, lymph nodes and bone marrow.
Histiocytic sarcoma	Histiocytic sarcoma – arising in red pulp expansion of large pale eosinophilic polymorphic cells often with indented nuclei. Giant cells may be present.	Hyaline droplets sometimes present in renal tubular epithelium. May also be seen in liver, uterus and other lymphoid organs.
Leukaemia	Granulocytic/myeloid leukaemia most common type in mice but less common than lymphoma. Spleen may have greenish colouration at necropsy. Red pulp expanded by monomorphic population of immature ring and band forms. Macrophages containing eosinophilic crystals may be present.	Similar changes in bone marrow. No obvious source of inflammation.
Blood vessel tumour	Irregular enlargement of spleen by irregular nodules of neoplastic blood vessels.	May be multicentric in mice so similar lesions may be seen in liver and mesenteric lymph nodes.

Source: Frith *et al.* (1996, 2001; Ward *et al.* (1999).

Figure 9.17　Extramedullary haematopoiesis in red pulp of the spleen characterized by the presence of megakaryocytes (black arrow) and cluster of small darkly basophilic staining erythrocyte precursors (white arrow).

Figure 9.18　Follicular/pleomorphic lymphoma of the spleen characterized by nodular expansion of splenic white pulp by pleomorphic lymphocytes of B cell origin.

megakaryocyte hyperplasia is seen in response to increased demand, as a result of anaemia (due to chronic haemorrhage for example from uterine and ovarian lesions in ageing female mice), neoplasia or inflammation (Figure 9.17). In severe cases of extramedullary haematopoiesis the spleen can be very large and the distinction between haematopoiesis and granulocytic leukaemia needs to be made based on the presence of an inciting lesion and the lack of involvement of other organs. Also, in leukaemia, a single stage of the developing granulocyte is typically

present whereas in reactive granulopoiesis a complete series of developing cells are usually present (Frith *et al.* 1983, 2007; Ward 1990). The white pulp is composed of three compartments: the PALS, the marginal zone, and follicles. The white pulp of the mouse tends to be more prominent than that of the rat, while the marginal zones are less prominent and more variable in the mouse. The PALS are divided into inner PALS and outer PALS, the inner being a T-cell dependant region stains more intensely then the outer PALS. The marginal zone is composed mainly of B-cells and macrophages, and a smaller population of T-cells. B-cell containing follicles are associated with the PALS, and show similar changes in morphology to those described in the lymph nodes (Cesta 2006a; Suttie 2006; Linden *et al.* 2012).

Other than extramedullary haematopoiesis, a variety of spontaneous changes are seen in the spleen of young and ageing mice. Accessory splenic tissue is occasionally seen in abdominal adipose tissue or the pancreas (Frith *et al.* 2007; Taylor 2011) (Figure 9.19) and atrophy of the red or white pulp occurs as a spontaneous change in old mice (Faccini *et al.* 1990; Suttie 2006) (Figure 9.20). Pigment accumulation, in the form of haemosiderin, can be seen in the red and white pulp of ageing mice, more commonly in females, and in pigmented

Figure 9.19　Accessory splenic tissue found as a discrete entity in the abdominal adipose tissues.

Figure 9.20   Reduced cellularity/atrophy of the white pulp of the spleen. Remnants of white pulp remain (arrows) but red pulp volume is maintained.

Figure 9.22   Haemangioma in the spleen of an aged CD-1 mouse.

mouse strains accumulation of melanin pigment can be identified (Suttie 2006; Taylor 2011). Special stains such as Perls' Prussian Blue Reaction for haemosiderin and Schmorl's stain for lipofuscin can be used to differentiate between the different pigments (Figure 9.21). Hyperplasia of the white pulp increases with age, with increases in number and size of the lymphoid components of the spleen. This is more common in female mice, although strain differences have been reported. Increases in the numbers of histiocytes, plasma cells and mast cells can also be seen in the red pulp of ageing mice (Frith *et al.* 2001; Bradley *et al.* 2012).

As with the lymph nodes, the spleen is often reported as a common site for deposition of amyloid,

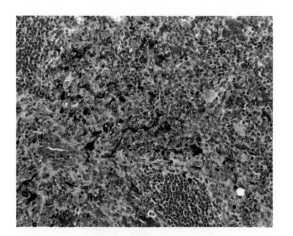

Figure 9.21   Melanin and haemosiderin pigments in the spleen of a pigmented mouse. The lighter pigment in the image stained positively with Perls' stain , the black pigment stained positively with Masson's Fontana.

although strain differences in incidence occur (Suttie 2006; Frith *et al.* 2007).

Lymphomas are commonly seen in the spleen in most mouse strains, and blood vessel tumours (haemangiomas and haemangiosarcomas) are reported in some strains of mice (Yamate *et al.* 1986; Faccini *et al.* 1990) (Figure 9.22).

9.4   MALT (mucosa-associated lymphoid tissue)

9.4.1   *Background and development*

The mucosa associated lymphoid tissue (MALT) represents dispersed aggregates of nonencapsulated lymphoid tissue along the surface of mucosal tissues and they act to protect the mucosae against infectious agents (Cesta 2006b; Brandtzaeg *et al.* 2008). Depending on the site, they are referred to variously as GALT (gut associated lymphoid tissue), BALT (bronchus-associated lymphoid tissue), NALT (nasal- – or nasopharynx- – associated lymphoid tissue), and about half the lymphocytes in the immune system are associated with these sites (Cesta 2006b).

At day 20 of embryonic development, areas of the gut can be recognized that have epithelial characteristics of Peyer's patches. Nasal-associated lymphoid tissue is visible before birth as accumulations of small groups of B-lymphocytes in the epithelium of the nasal floor. T-lymphocytes appear after birth. The number of B- and T-cells increases postnatally to the size of adult NALT by the age of 4 weeks; BALT develops postnatally (van Rees *et al.* 1996).

9.4.2   *Sampling techniques*

Routinely, transverse sections of the jejunum at the sites of grossly obvious Peyer's patches are taken, but utilizing the Swiss-roll technique can allow evaluation of the entire intestine with associated GALT (Ruehl-Fehlert *et al.* 2003).

Varying amounts of NALT are visible on the routine sections of the nasal turbinates presented for histopathological evaluation in toxicity studies. Paired aggregates of lymphoid tissue are present at the entrance to the nasopharyngeal duct, throughout routine level III of the standard trimming levels for the mouse head (Herbert and Leininger 1999; Kittel *et al.* 2004).

The presence of BALT in the lungs of mice varies with strain, and housing conditions, and does not develop without antigenic stimulation (Cesta 2006b), but trimming the lobes of the lung in longitudinal sections to allow presentation of the lobar bronchus and its main branches will allow the best opportunity for identifying the presence of BALT (Kittel *et al.* 2004).

9.4.3   *Artefacts*

Consistency of sampling will ensure that variation in the presentation and size of MALT between animals is kept to a minimum. The lymphoid tissue of the GALT is prone to crush artefacts if handled incorrectly at necropsy or histology, similar to those seen in lymph nodes (Figure 9.4).

9.4.4   *Anatomy and histology*

The structure of MALT resembles that of lymph nodes, with variable numbers of B-cell rich follicles, and intervening T-cell zones. They also contain macrophages and antigen presenting cells and

high endothelial vessels are present. Unlike lymph nodes, however, they do not have afferent lymphatics and antigens are sampled directly across mucosal surfaces. The epithelium overlying MALT is specialized and is composed of 'microfold' or 'membrane' M cells (Brandtzaeg *et al.* 2008, Cesta 2006b). The GALT is characterized by the presence of discrete collections of follicles (Peyer's patches) and isolated lymphoid follicles located within the mucosa of the small intestine, as well as isolated lymphoid follicles and lymphoid aggregates in the large intestine (Figures 3.27 and 3.28). Peyer's patches occur with the greatest frequency in the jejunum, on the antimesenteric border (Cesta 2006b; Brandtzaeg *et al.* 2008). Mice do not have tonsils (Cesta 2006b; Brandtzaeg *et al.* 2008; Linden *et al.* 2012). Bronchus-associated lymphoid tissue is the main lymphoid organ of the lungs and when present is found along the main stem and branches of the bronchi. Nasal-associated lymphoid tissue is visible as paired aggregates of lymphoid tissue at the entrance to the nasopharyngeal duct, and distinct B- and T-cell areas are evident (van Rees *et al.* 1996).

Antigenic stimulation leads to an increase of MALT and increases or decreases in the size or cell density of MALT are the most commonly reported histopathological changes, although the term 'lymphoid hyperplasia' is usually reserved for a prominent increase in lymphoid tissue in an area where MALT is not usually conspicuous (Taylor 2011) (Figure 9.23).

9.5   Thymus

9.5.1   *Background and development*

The thymus is a primary immune organ, in which T-cells are produced without the need for antigenic stimulation. It is essential for normal development and function of the immune system (Ward *et al.* 1999).

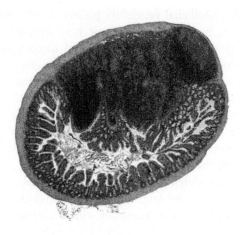

Figure 9.23   Lymphoid hyperplasia of MALT.

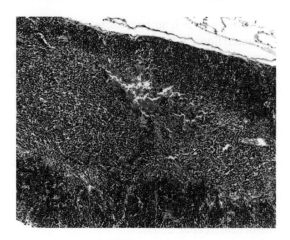

Figure 9.24   Haemorrhage of the thymus, probably arte-factual as a result of damage to the tissue during necropsy procedures.

The thymus develops from around embryonic day 12, from two separate primordia, the endoderm of the third and fourth pharyngeal pouches, and adjacent mesenchyme. The primordia migrate caudally into the thorax and fuse during days 14 and 15, and from then develop into an adult-like structure, reaching adult size at sexual maturity after which it gradually involutes (van Rees *et al.* 1996; Pearse 2006a).

9.5.2   Sampling techniques

The size and weight of the mouse thymus reflects changes in immune status of the organ in response to a number of factors, including age, sex, strain of animal and environmental factors such as diet and stress (Elmore 2006b; Pearse 2006b). Consistent sampling procedures ensure comparability between animals. Routinely, the thymus is trimmed along the length of one lobe but where investigation of immuno-toxicological effects is required, it is better to embed the whole organ, and to section both lobes (Morawietz *et al.* 2004; Pearse 2006a).

9.5.3   Artefacts

Incorrect orientation of the thymus in the block can lead to tangential sections, which show artefactual variations in cortical:medullary ratios (Elmore 2012). Artefactual haemorrhage can also be seen as an agonal change, or following damage during necropsy procedures (Figure 9.24).

9.5.4   Anatomy and histology

The thymus is a bi-lobed organ located in the anterior thoracic cavity in the pericardial mediastinum. The two lobes are connected by connective tissue but there is no parenchymal connection between the two lobes. Connective tissue forms the capsule, and connective tissue septae divide the lobe into several lobules. The parenchyma consists of a network of epithelial cells, which is populated by lymphocytes, dendritic cells and macrophages. Each lobule displays similar features with a darkly staining cortex surrounding a paler staining medulla (Figure 9.25). The cortex is packed with developing T-cells, and the medulla contains larger, paler staining lymphocytes. Thymic arteries follow the course of the connective tissue septa and branch to form capillary arcades at the corticomedullary junction. The epithelial components are more obvious in the medulla, but Hassall's corpuscles (eosinophilic concentric whorls of flattened epithelial cells) are not as prominent in the mouse as in other species (van Rees *et al.* 1996; Pearse 2006a). The

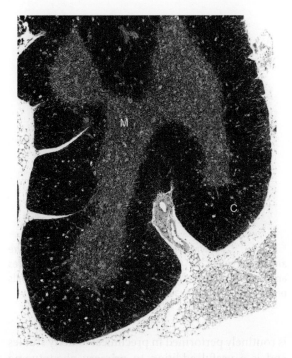

Figure 9.25 Normal structure of thymus – featuring distinction between cortex (C) and medulla (M).

thymus of females is usually larger than in males but thymic involution – the normal age-related decrease in cellularity of the thymus – begins after sexual maturity (Figure 9.26). There are strain and sex differences in the timing and rate of thymic involution but it is characterized by a reduction in the size of the thymus, and reduced cellularity of the different regions of the thymus leading to a

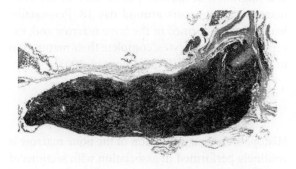

Figure 9.26 In thymic involution and atrophy there is overall loss of tissue and a loss of distinction between cortex and medulla.

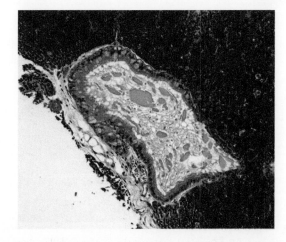

Figure 9.27 Cystic structure within the thymic parenchyma.

loss of corticomedullary demarcation; however, the thymus of the mouse does not completely involute (Elmore 2006b; Pearse 2006b; Gui *et al.* 2012).

Some of the most common spontaneous changes involving the thymus are ectopic thymus in the region of the thyroids and parathyroids (Figure 7.22), and epithelial cysts within the thymus (Figure 9.27). Ectopic parathyroid tissue within the thymus is also rarely seen. These changes are due to the close association of the thyroid and parathyroid anlage with the thymic anlage during embryonic development (Dooley *et al.* 2006; Pearse 2006a,b; Taylor 2011). Germinal centres can be seen at the corticomedullary junction of aged mice (Taylor 2011) (Figure 9.28). Lymphoid hyperplasia is also commonly seen in the thymus of aged mice and can be seen in association with thymic involution / atrophy (Taylor 2011) (Figure 9.29).

The occurrence of thymic lymphoma is common and thymomas also occur infrequently. The term 'thymoma' is used to describe lesions in the thymus characterized by the presence of a neoplastic epithelial component, considered to be derived from the epithelial cells of the reticular framework of the thymus. They are often solitary, well demarcated masses composed of varying proportions of lymphocytes and thymic epithelial cells (Frith *et al.* 2001 Figure 9.30).

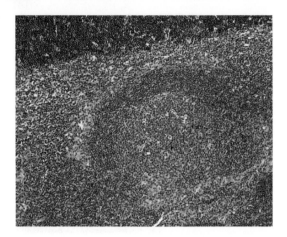

Figure 9.28   Germinal centre at the corticomedullary junction of the thymus in an aged mouse.

Figure 9.30   Thymoma with loss of normal architecture and replacement by varying proportions of lymphocytes and thymic epithelial cells. Pale staining epithelial cells predominate in this example.

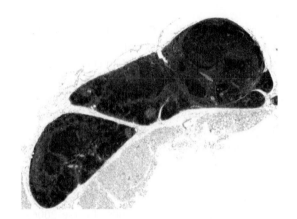

Figure 9.29   Lymphoid hyperplasia of the thymus in an aged mouse. Remnants of cortex and medulla are still evident. Nodules of hyperplastic lymphocytes with germinal centres distinguish this from a neoplastic infiltrate.

9.6   Bone marrow

9.6.1   *Background and development*

The bone marrow is the primary site for haematopoiesis and is a primary lymphoid tissue. It is located in the hollow spaces of bones, and examination of the bone marrow in formalin-fixed, H&E stained sections of bone (following decalcification) is routinely performed in preclinical toxicity studies and is a useful addition to primary phenotyping screens.

The cells of the bone marrow originate from pluripotential haematopoietic stem cells, which develop in the mouse embryo from embryonic day 8 onwards, and join the embryonic circulation. They migrate to the foetal liver at day 10, where they differentiate to form subpopulations of stem cells for lymphoid and myeloid cell lineages. Haematopoietic stem cells are found in the foetal spleen by day 13. The formation of the bone marrow cavity occurs late in gestation, as the bones are mineralized, and migration of haematopoietic cells to the bone marrow cavity occurs around day 18. Postnatally, leukocytes are formed in the bone marrow and, except for T–lymphocytes, complete their maturation in the bone marrow (Holsapple *et al.* 2003).

9.6.2   *Sampling techniques*

Histopathological evaluation of the bone marrow is routinely performed in association with sections of sternum and femur which have previously been fixed in formalin and undergone decalcification. A longitudinal section of the knee joint, including the distal femur and proximal tibia, and a longitudinal horizontal section of the sternum are recommended for evaluation of bone and bone marrow morphology (Morawietz *et al.* 2004).

The examination of bone marrow smears, or flow cytometric evaluation of the bone marrow, is not covered in this chapter. The reader is referred to Travlos (2006a) and the best practice document (Reagan *et al.* 2011) for further information on the best use of these techniques.

9.6.3 *Artefacts*

Artefactual shrinkage of the bone marrow associated with the histologic processing of the bone is common (Travlos 2006a; McInnes 2011) and care must be taken in distinguishing degenerative changes of the bone marrow from processing induced artefactual changes (Travlos 2006b).

9.6.4 *Anatomy and histology*

The bone marrow is composed of a loose connective tissue stroma, which is highly vascularized, and is populated by the haematopoietic cells. The bony cavities are normally filled with active marrow, composed of varying populations of haematopoietic cells, including granulocytic, erythrocytic and megakaryocytic elements and stem cells (Figure 9.31a). An estimate of general haematopoietic activity and myeloid:erythroid ratio can be made on H&E-stained sections (Faccini *et al.* 1990; Travlos 2006a). Different components of the bone marrow can be identified. The megakaryocytes are large cells with multinucleated nuclei; erythroid cells are small with round, deeply basophilic nuclei; and granulocytes are less basophilic than erythropoietic cells, with bean-shaped nuclei (Travlos 2006a).

Focal degeneration or atrophy of the bone marrow is occasionally seen as a spontaneous change in mice (Travlos 2006b) but fatty infiltration of the bone marrow is usually limited (Anver and Haines 2004). There are strain differences in the occurrence of fibro-osseous lesion, which occurs as a spontaneous lesion in the bone marrow of the sternum

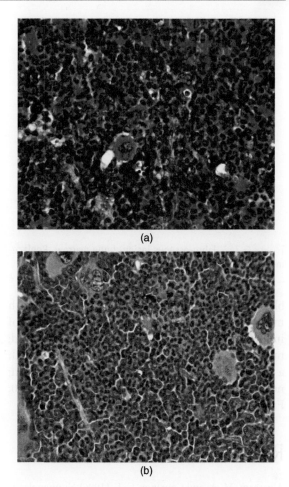

(a)

(b)

Figure 9.31 (a) Bone marrow, illustrating normal haematopoietic elements. (b) Myeloid hyperplasia.

and long bones (Figure 9.32) and it is generally more common in females than males (Albassam and Courtney 1996; Travlos 2006b). Severe granulocytic hyperplasia (myeloid hyperplasia) can occur secondary to inflammatory changes in the mouse, and these need to be distinguished from myeloid leukaemia (Figure 9.31b). Features to take account of in this interpretation include the following. In reactive extramedullary haematopoiesis there are likely to be mature cells in the peripheral blood; there is usually minimal enlargement of the liver and lymph nodes; an inciting stimulus is usually apparent (for example, an abscess, neoplastic growth, increased

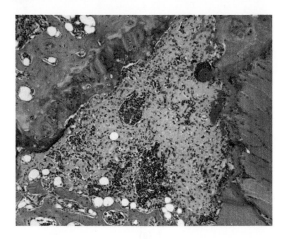

Figure 9.32 Fibro-osseous lesion, bone marrow. Replacement of bone marrow with pale eosinophilic material.

blood loss). In contrast, features associated with myeloid leukaemia include an increased number of immature precursor cells in the peripheral blood; marked enlargement of liver and lymph nodes with a pronounced uniform leukaemic cell infiltration; no apparent inciting stimulus (Faccini *et al.* 1990; Frith *et al.* 2001; Travlos 2006b).

References

Albassam, M. A. and Courtney, C. L. (1996) Nonneoplastic and neoplastic lesions of the bone in *Pathobiology of the Aging Mouse*, Vol 2. (eds U. Mohr, D. L. Dungworth, C. C. Capen, *et al.*), ILSI Press, Washington, pp. 425–437.

Anver, M.R. and Haines, D.C. (2004) Gerontology, in *The Handbook of Experimental Animals: The Laboratory Mouse* (eds H.J. Hedrich and G. Bullock), Elsevier, San Diego, CA, pp. 327–343.

Bradley, A., Mukaratirwa, S. and Petersen-Jones, M. (2012) Incidences and range of spontaneous findings in the lymphoid and haemopoietic system of control Charles River CD-1 mice (Crl: CD-1(ICR) BR) used in chronic toxicity studies. *Toxicologic Pathology* **40**, 375–381.

Brandtzaeg, P., Kiyono, H., Pabst, R. and Russell, M.W. (2008) Terminology: nomenclature of mucosa-associated lymphoid tissue. *Mucosal Immunology* **1**, 31–37.

Cesta, M.F. (2006a) Normal structure, function, and histology of the spleen. *Toxicologic Pathology* **34**, 455–465.

Cesta, M.F. (2006b) Normal structure, function, and histology of mucosa-associated lymphoid tissue. *Toxicologic Pathology* **34**, 599–608.

Cook, M.J. (1965) *The Anatomy of the Laboratory Mouse*, Academic Press, London, http://www.informatics.jax.org/cookbook/ (accessed 19 July 2013).

Cook, M.J. (1983) Anatomy, in *The Mouse in Biomedical Research*, Vol III, Normative Biology, Immunology, and Husbandry(eds H.L. Foster, J.D. Small, and J.G. Fox), Academic Press Inc, Boston, MA, pp 101–120.

Dooley, J., Erickson, M., Gillard G.O. and Farr, A.G. (2006) Cervical thymus in the mouse. *Journal of Immunology* **176**, 6484–6490.

Elmore, S.A. (2006a) Histopathology of the lymph nodes. *Toxicologic Pathology* **34**, 425–454.

Elmore, S.A. (2006b) Enhanced histopathology of the thymus. *Toxicologic Pathology* **34**, 656–665.

Elmore, S.A. (2012) Enhanced histopathology of the immune system: a review and update. *Toxicologic Pathology* **40**, 148–156.

Engelhardt, J.A., Gries, C.L. and Long, G.G. (1993) Incidence of spontaneous neoplastic and nonneoplastic lesions in Charles River CD-1 mice varies with breeding origin. *Toxicologic Pathology* **21**, 538–541.

Faccini, J.M., Abbott, D.P. and Paulus, G.J.J. (1990) *Mouse Histopathology*, Elsevier, Amsterdam.

Frith, C.H., Goodman, D.G. and Boysen, B.G. (2007) The mouse, pathology, in *Animal Models in Toxicology*, 2nd edn (ed. S.C. Gad), Taylor & Francis, Boca Raton, FL, pp. 72–122.

Frith, C.H., Highman, B., Burger, G., and Sheldon, W.D. (1983) Spontaneous lesions in virgin and retired breeder BALB/c and C57BL/6 mice. *Laboratory Animal Science* **33**, 273–286.

Frith, C.H., Ward, J.M., Frederickson, T. and Harleman, J.H. (1996) Neoplastic lesions of the hematopoietic system, in *Pathobiology of the Aging Mouse, Vol 1*, (eds U. Mohr, D.L. Dungworth, C.C. Capen *et al.*), ILSI Press, Washington, pp. 219–235.

Frith, C.H., Ward, J.M., Harleman, J.H. *et al.* (2001) Hematopoietic system, in *International Classification of Rodent Tumors, The Mouse* (ed U. Mohr), Springer, Berlin, pp. 417–451.

Gui, J., Mustachio, L.M., Su, D.M. and Craig, R.W. (2012) Thymus size and age-related thymic involution: early programming, sexual dimorphism, progenitors and stroma. *Aging and Disease* **3**, 280–290.

Haley, P. (2003) Species differences in the structure and function of the immune system. *Toxicology* **188**, 49–71.

Haley, P., Perry, R., Ennulat, D. *et al.* (2005) STP position paper: best practice guideline for the routine pathology evaluation of the immune system. *Toxicologic Pathology* **33**, 404–407.

Herbert, R.A. and Leininger, J.R. (1999) Nose, larynx, and trachea, in *Pathology of the Mouse, Reference and Atlas.* (eds R.R. Maronpot, G.A. Boorman and B.W. Gaul) Cache River Press, Vienna, pp. 259–292.

Holsapple, M.P., West, L.J. and Landreth, K.S. (2003) Species comparison of anatomical and functional immune system development. *Birth Defects Research. Part B, Developmental and Reproductive Toxicology* **68**, 321–334.

Hummel, K.P., Richardson, F.L. and Fekete, E. (1966) Anatomy, in *Biology of the Laboratory Mouse* (ed E.L. Green), McGraw-Hill, New York, pp. 247–307, http://www.informatics.jax.org/greenbook/frames/frame13.shtml (accessed 19 July 2013).

Kittel, B., Ruehl-Fehlert, C., Morawietz, G. *et al.* (2004) Revised guides for organ sampling and trimming in rats and mice – Part 2. *Experiments in Toxicologic Pathology* **55**, 413–431, http://reni.item.fraunhofer.de/reni/trimming/index.php (accessed 19 July 2013).

Linden, M., Ward, J.M. and Cherian, S. (2012) Hematopoietic and lymphoid tissues, in *Comparative Anatomy and Histology: A Mouse and Human Atlas* (eds P.M. Treuting and S.M. Dintzis), Elsevier Inc., Amsterdam, pp 309–338.

Maita, K., Hirano, M., Harada, T., *et al.* (1988) Mortality, major cause of moribundity, and spontaneous tumors in CD-1 mice. *Toxicologic Pathology* **16**, 340–349.

Maronpot, R.R. (2006) A monograph on histomorphologic evaluation of lymphoid organs. *Toxicologic Pathology* **34**, 407–408.

McInnes, E.F. (2011) Artifacts in histopathology, in *Background Lesions in Laboratory Animals. A Color Atlas* (ed. E.F. McInnes), Saunders, Edinburgh, pp. 93–99.

Mikaelian, I., Nanney, L.B., Parman, K.S. *et al.* (2004) Antibodies that label paraffin-embedded mouse tissues: a collaborative endeavor. *Toxicologic Pathology* **32**, 181–191.

Morawietz, G., Ruehl-Fehlert, C., Kittel, B., *et al.* (2004) Revised guides for organ sampling and trimming in rats and mice – part 3. *Experimental Toxicologic Pathology* **55**, 433–449, http://reni.item.fraunhofer.de/reni/trimming/index.php (accessed 19 July 2013).

Pearse, G. (2006a) Normal structure, function and histology of the thymus. *Toxicologic Pathology* **34**, 504–514.

Pearse, G. (2006b) Histopathology of the thymus. *Toxicologic Pathology* **34**, 515–547.

Percy, D.H. and Barthold, S.W. (2007) Mouse, in *Pathology of Laboratory Rodents and Rabbits*, 3rd edn, Blackwell Publishing, Ames, IA, pp. 3–124.

Reagan, W.J., Irizarry-Rovira, A., Poitout-Belissent, F. *et al.* (2011) Best practices for evaluation of bone marrow in nonclinical toxicity studies. *Toxicologic Pathology* **39**, 435–448.

Rehg, J.E., Bush, D. and Ward, J.M. (2012) The utility of immunohistochemistry for the identification of hematopoietic and lymphoid cells in normal tissues and interpretation of proliferative and inflammatory lesions of mice and rats. *Toxicologic Pathology* **40**, 345–374.

Ruehl-Fehlert, C., Kittel, B., Morawietz, G. *et al.* (2003) Revised guides for organ sampling and trimming in rats and mice – Part 1. *Experimental Toxicologic Pathology* **55**, 91–106, http://reni.item.fraunhofer.de/reni/trimming/index.php (accessed 19 July 2013).

Sellers, R.S., Clifford, C.B., Treuting, P.M. and Brayton, C. (2012) Immunological variation between inbred laboratory mouse strains: points to consider in phenotyping genetically immunomodified mice. *Veterinary Pathology* **49**, 32–43.

Seymour, R., Ichiki, T., Mikaelian, I. *et al.* (2004) Necropsy methods, in *The Handbook of Experimental Animals: The Laboratory Mouse* (eds H.J. Hedrich and G. Bullock), Elsevier, San Diego, CA, pp. 495–516.

Suttie, A.W. (2006) Histopathology of the spleen. *Toxicologic Pathology* **34**, 466–503.

Taylor, I. (2011) Mouse, in *Background Lesions in Laboratory Animals.* A Color Atlas (ed. E.F. McInnes), Saunders, Edinburgh, pp. 45–72.

Travlos, G.S. (2006a) Normal structure, function, and histology of the bone marrow. *Toxicologic Pathology* **34**, 548–565.

Travlos, G.S. (2006b) Histopathology of bone marrow. *Toxicologic Pathology* **34**, 566–598.

Van den Broeck, W., Derore, A. and Simoens, P. (2006) Anatomy and nomenclature of murine lymph nodes: descriptive study and nomenclatory standardization in BALB/cAnNCrl mice. *Journal of Immunological Methods* **312**, 12–19.

van Rees, E.P., Sminia, T., Dijkstra, C.D. (1996) Structure and development of the lymphoid organs, in *Pathobiology of the Aging Mouse, Vol 1*, (eds U. Mohr, D.L. Dungworth, C.C. Capen, W.W. Carlton, J.P. Sundberg and J.M. Ward), ILSI Press, Washington, pp. 173–187.

Ward, J.M. (1990) Classification of reactive lesions, spleen, in *Hemopoietic System, Monographs on Pathology of Laboratory Animals* (eds T.C. Jones, J.M. Ward, U. Mohr and R.D. Hunt), Springer-Verlag, Berlin, pp. 220–226.

Ward, J.M., Erexson, C.R., Faucette, L.J. *et al.* (2006) Immunohistochemical markers for the rodent immune system. *Toxicologic Pathology* **34**, 616–630.

Ward, J.M., Mann, P.C., Morishima, H. and Frith, C.H. (1999) Thymus, spleen, and lymph nodes, in *Pathology of the Mouse, Reference and Atlas.* (eds R.R. Maronpot, G.A. Boorman and B.W. Gaul), Cache River Press, Vienna, pp. 333–360.

Ward, J.M., Rehg, J.E. and Morse, H.C. (2012) Differentiation of rodent immune and hematopoietic system reactive lesions from neoplasias. *Toxicologic Pathology* **40**, 425–434.

Willard-Mack, C.L. (2006) Normal structure, function, and histology of lymph nodes. *Toxicologic Pathology* **34**, 409–424.

Yamate, J., Tajima, M. and Kudow, S. (1986) Comparison of age-related changes between long-lived CRJ:CD-1 (ICR) and CRJ:B6C3F1 mice. *Nihon Juigaku Zasshi* **48**, 273–284.

Chapter 10
Integument and adipose tissue

Cheryl L. Scudamore
Mary Lyon Centre, MRC Harwell, UK

10.1 Background and development

The integumentary system in the mouse is similar to that of other mammals, consisting of the skin and its associated specialized structures (mammary tissue, hair and nails). Components of the skin are derived from multiple embryonic layers with the epidermis deriving from ectoderm, connective tissues of the dermis and subcutis from mesoderm and nerve endings and melanocytes from the neural crest. The epidermis develops from a single layer in the embryo to become a multilayered structure which is thickest at birth and then thins to two to three layers in the first two weeks postnatal. The first hair follicle cycle of development also begins in the embryo but does not complete until after birth. Different types of hair follicle develop at different rates with the vibrissae being most advanced at birth (Sundberg and King 2000).

In the female mouse the mammary gland arises from ectoderm around embryonic day 11, proliferating to invade the mammary fat pad around day 17 and then forming a simple branching structure with a single duct leading to the nipple prior to birth. In the male embryo development of the mammary epithelium is curtailed at day 13 leaving only remnant tubular structures within the mammary fat. The majority of mammary development occurs postnatally in the female with the onset of ovarian hormone production from week 3 (Richert *et al.* 2000).

The origins of adipocytes have been under intense investigation in recent years and, while most adipose tissue is still thought to arise from mesoderm, there is evidence of neurectodermal origin for some fat deposits and postnatal development from bone-marrow stem cells. White adipocytes and brown adipocytes also appear to arise from different lineages with brown adipocytes expressing genes suggestive of a myogenic cell origin (Majka *et al.* 2011).

A Practical Guide to the Histology of the Mouse, First Edition. Cheryl L. Scudamore.
© 2014 John Wiley & Sons, Ltd. Illustrations © Veterinary Path Illustrations, unless stated otherwise.
Published 2014 by John Wiley & Sons, Ltd. Companion Website: www.wiley.com/go/scudamore/mousehistology

10.2   Sampling technique

For routine sampling neutral buffered formalin is sufficient for fixation of skin, mammary and adipose tissue for morphological analysis, although acid alcohol fixatives and freezing may be needed for immunohistochemistry (Paus *et al.* 1999; Sundberg and King 2000). The site of sampling may reflect the study requirements. A single section from the inguinal region will enable analysis of the skin, mammary and white adipose tissue (including nipple in female mice) (Ruehl-Fehlert *et al.* 2003) in one sample. Mammary tissue is extensive in the female mouse (Chapter 1, Figure 1.14) and so may also be seen incidentally in skin samples from multiple sites.

For more detailed analysis of hair-follicle development, a sample of dorsal skin is preferred. Ideally the samples should be taken from a consistent site, for example in the dorsal midline at the thoracolumbar junction (Paus *et al.* 1999). To aid orientation after fixation, samples of skin can be laid flat onto cardboard at the time of sampling to prevent distortion (Paus *et al.* 1999; Sundberg and King 2000). Labelling the cranial and caudal edges of the sample will also help ensure that samples are sectioned to produce longitudinal sections through the hair follicles which are easier to analyse than cross or tangential sections. Although H&E staining is adequate for basic analysis of skin and hair follicles, other histochemical and immunohistochemical stains may be useful to delineate accurately all the substages of hair follicle development (Muller-Rover *et al.* 2001).

A single skin sample taken from mice at 10–12 weeks of age may be suitable for a routine screen where no skin or hair phenotype is suspected. Where a phenotype is suspected then it is important to sample animals at different stages of the hair developmental cycle (Sundberg *et al.* 2005) and also to analyse plucked hair samples. In depth approaches to phenotyping have been described in detail elsewhere and will not be covered in this chapter (Sundberg and King 2000).

Additional samples of skin may be needed to examine skin of different types (e.g. ear, eyelid, muzzle, tail) and nails. Muzzle or facial skin samples may be taken for analysis of vibrissae. Tail skin is most easily examined by fixation of the whole tail and decalcification prior to sectioning. Nails can be sectioned longitudinally after dissection, fixation and decalcification of individual digits. Sampling of pinna and eyelids are discussed in Chapter 12.

Analysis of mammary tissue from models of mammary tumours may require more extensive tissue sampling of specific glands including representative thoracic and inguinal mammary glands in addition to any tumour masses (Cardiff *et al.* 2000). Trimming of excess fat from the mammary tissue will aid fixation and processing. At necropsy, lesions can be noted using a map of mammary tissue (Rasmussen *et al.* 2000). For detailed analysis of the branching structure of the mammary glands, whole mount sections may be useful (Rasmussen *et al.* 2000).

Until recently, adipose tissue was routinely divided into two main subtypes – brown adipose tissue (BAT) and white adipose tissue (WAT), based on their morphological appearance and the ability of BAT to express UCP1. It is now accepted that a third subtype known as 'brite' adipose tissue consists of adipose tissue that is WAT in morphological appearance but can also express UCP1 when stimulated, for example by exposure to cold. In routine studies, adipose tissues may not be sampled specifically but their presence may be noted in association with other tissue samples. For example BAT is often found surrounding the aorta in mice (Chapter 4) and WAT is often seen subcutaneously in skin or surrounding mammary tissue. Where specific investigation of adipose tissue is required, for example in the study of metabolic and obese phenotypes, appropriate deposits can be sampled (Casteilla *et al.* 2008) to take into account BAT, WAT and brite phenotypes (Walden *et al.* 2012). In the mouse the interscapular, cervical, axillary and mediastinal fat deposits all express UCP-1. The bi-lobed interscapular deposit is the most well developed and is a convenient site to sample in all ages of mice. Cardiac, inguinal and retoperitoneal fat deposits can express UCP-1 under certain conditions and are classified as 'brite' adipose tissue. Brite tissues are rarely sampled but it is important to be aware of this mixed phenotype. Mesenteric and epididymal do not normally express UCP-1 and can be sampled as examples of

WAT (Casteilla *et al.* 2008). Tissues from specific fat deposits should be embedded in a consistent way. It is sometimes useful to take a cross-section through the interscapular fat at the widest point as this allows analysis of the size and consistency of the BAT and the sample also usually includes some peripheral WAT for ease of comparison.

10.3 Anatomy and histology

The histology of mouse skin is generally similar to that of other mammals, consisting of an outer epithelial epidermis covering the connective tissue dermis and an underlying hypodermis largely composed of fat (Figure 10.1). A thin layer of skeletal muscle, the panniculus carnosus, separates the hypodermal fat from the adventitia, a loose connective tissue layer attached to the muscle of the underlying body wall.

The epidermis may appear very thin (1–2 layers) in adult mice but still consists of the four epithelial-derived cell layers seen in other species – stratum basale, stratum spinosum and stratum granulosum covered by the keratinized stratum corneum (Figure 10.2), which can be distinguished by different patterns of keratin expression using IHC. The stratum basale contains the proliferating cells and so is the only layer where mitoses may be seen in normal skin. Darkly basophilic granules may be seen in the stratum granulosum cells. The epidermis is thicker in neonatal mice (Figure 10.3) and the hair follicles are still in an embryonic hair cycle. The

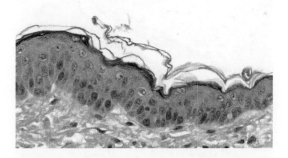

Figure 10.2 High-power image of hyperplastic epidermis. The epidermis in this section is slightly thicker than normal but allows all the layers to be seen, including the stratum granulosum, which can be hard to visualize in normal skin sections.

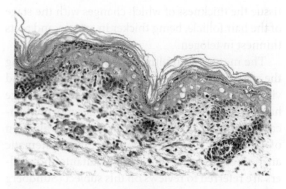

Figure 10.3 Embryonic and neonatal skin has more visible layers than normal adult skin.

first mature hair cycle starts 2 to 3 weeks after birth. In pigmented strains of mice, melanin pigment is found in the cells of the stratum basale.

The dermis consists of loose connective tissue with blood vessels and normally a sparse resident population of inflammatory cells. Hair follicles and associated adnexal glands and muscle are also seen in the dermis and extending into the hypodermis. The sebaceous glands are made up of large vacuolated polygonal cells with a central nucleus. The cytoplasm of these secretory cells can be stained for neutral lipid, for example with oil-red-O. The secretory cells are surrounded by a population of small, cuboidal, darkly staining cells known as reserve cells which differentiate into secretory cells (Figure 10.4). The hypodermis is composed predominantly of adipose

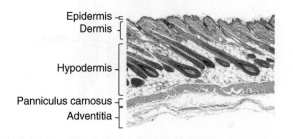

Epidermis
Dermis

Hypodermis

Panniculus carnosus
Adventitia

Figure 10.1 Overview of skin layers and anagen hair follicles.

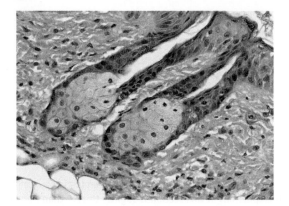

Figure 10.4   Sebaceous glands composed of vacuolated secretory cells surrounded by small reserve cells.

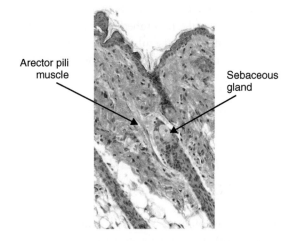

Figure 10.5   Superficial portion of the follicle includes the isthmus and infundibulum differentiated by the positions of the sebaceous glands and arector pili muscle.

tissue the thickness of which changes with the stage of the hair follicle, being thicker in anagen and at its thinnest in telogen.

The mouse has multiple types of hair, including the specialized hairs that make up the eyelashes and vibrissae. The trunk is covered in truncal or pelage hairs of several subtypes, which differ morphologically from tail, ear and perigenital hairs. Evaluation of these hair types is important for a systematic analysis of suspected hair follicle mutants and readers are referred to reviews on this subject (Sundberg *et al.* 2005). It is important to be aware of the cyclicity of hair follicle development and that this changes with age. The hair follicle cycles through a series of major stages (anagen, catagen, telogen and exogen), which, inevitably, are made up of multiple intermediary stages (Paus *et al.* 1999, Muller-Rover *et al.* 2001). For routine phenotyping and toxicological studies it is important that a consistent skin site is sampled (this minimizes the potential differences due to waves of hair development). This is also essential in order to recognize the major hair follicle stages, as it is easy for the inexperienced to confuse follicles at different stages of the hair cycle with a potential phenotype.

The basic structure of the hair follicle is similar for all hair types and can be divided into four regions. The infundibulum is the region between the surface epithelium and the entrance to the sebaceous glands, the isthmus between the sebaceous gland and

the insertion of the arrector pili muscle (formed of smooth muscle) (Figure 10.5), the lower follicle between the muscle and the hair bulb and the hair bulb itself (Figure 10.6). Anagen hairs are in the growing phase and are the largest and most easily recognized due to the conspicuous hair bulb seen in the deep layers of the skin (Figure 10.1 and Figure 10.6). The actual length of the follicle depends on the hair type. The bulb consists of outer layers of epithelial

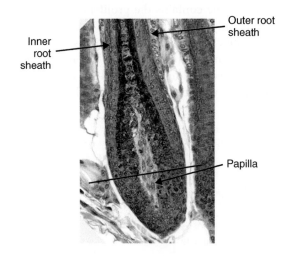

Figure 10.6   Detail of anagen hair bulb with Auber's line shown by dark line.

cells surrounding a central core of modified fibroblasts and capillaries – the dermal papilla. The dermal papilla can be stained with histochemical stains for alkaline phosphatase during all stages of the hair cycle. The matrix epithelial cells surrounding the dermal papilla differentiate to form the hair and the layers of the hair shaft. The hair follicle is surrounded by a basement membrane (glassy membrane), which is adjacent to the outer root sheath epithelial layer and the inner root sheath layer, which is adjacent to the hair shaft. Pigment is often seen in the epithelial cells around the dermal papilla arising abruptly at the middle of the hair bulb at Auber's line (Figure 10.6).

In the catagen phase there is a wave of apoptosis and the hair follicles start to regress. Stains for apoptosis can help to distinguish this stage. This phase of the hair cycle takes 2–4 days. The follicles are shorter than in anagen, with a less conspicuous bulb and loss of pigment (Figure 10.7). The glassy membrane may be thickened and prominent and the club hair is formed.

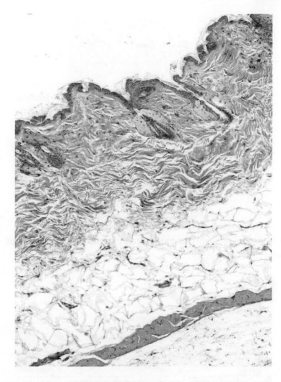

Figure 10.8   Telogen hair follicles.

Telogen is the resting phase of the hair cycle when the dermal papilla surrounded by a thin layer of epithelial cells retracts to its shortest length with the papilla coming to lie at the level of the sebaceous glands in the dermis (Figure 10.8). The dermal papilla is round in shape and there is no obvious inner root sheath.

Vibrissae are specialized hairs found on the muzzle and lower legs of mice. The structure of the vibrissae is essentially similar to other hair follicles in that they consist of an epithelial core consisting of the outer root sheath, inner root sheath and hair surrounded by a mesenchymal sheath (glassy membrane). The vibrissae differ from other pelage hairs by the presence of a vascular sinus surrounding the epithelial core, the appearance of which can be confused with a vascular abnormality by the inexperienced (Figure 10.9). The vascular sinus is contained within a thick fibrous capsule. Skeletal muscle fibres, which control movement and sensory nerves, are associated with the capsule (Dorfl 1982).

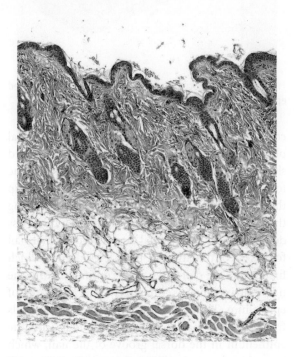

Figure 10.7   Catagen hair follicles.

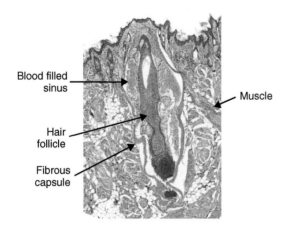

Figure 10.9   Vibrissa hair surrounded by blood-filled sinus and fibrous capsule.

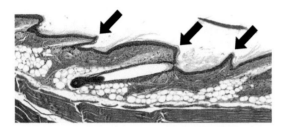

Figure 10.10   Tail skin with thicker epidermis and thick hair shafts leading to a scale-like appearance of the surface (arrows).

The hair follicles and hairs of the tail are shorter and thicker than on the trunk and the epidermis is thicker and more heavily keratinized (Figure 10.10).

The epidermis is also thicker and more heavily keratinized over the mouse foot pad. This is also the only anatomic region of mouse skin where eccrine sweat glands are found. The eccrine glands are made up of coiled acini, lined by cuboidal epithelium (Figure 10.11), which empty through the eccrine duct that runs through the thickened epidermis of the foot pad.

10.3.1   *Nail*

The mouse nail is structurally similar to human nail with a hard nail plate made of multiple keratin types and subtypes (types 5, 6, 14, 16, 17, Wong

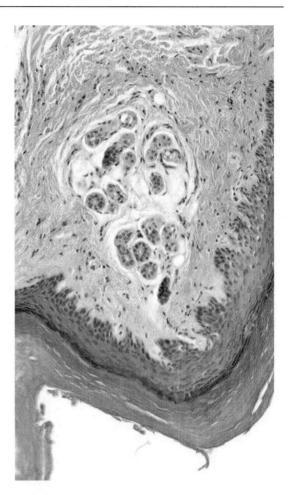

Figure 10.11   Footpad has thick and heavily keratinized epidermis with eccrine sweat glands in the dermis.

et al. 2005) overlying the dorsal surface of the distal phalanx. The nail arises abruptly from the haired skin of the foot dorsally and the modified skin of the foot pad ventrally. The epithelium, which makes up the nail structure, is similar to that of skin having the same four basic layers, which differ in thickness depending on their actual site. The proximal nail fold is a continuation of the toe epidermis which first covers the nail and is then continuous with the nail matrix which is a relatively thickened area of epithelium. The epithelium of the nail bed extends from the nail matrix and is continuous with the epithelium in the dorsal groove of the nail, which has a thick keratinized layer that is continuous with

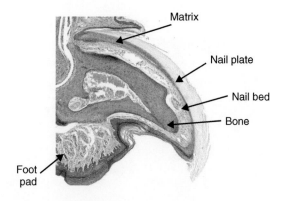

Figure 10.12   Nail.

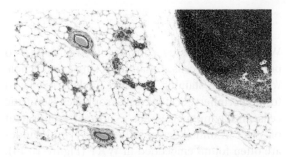

Figure 10.14   More extensive mammary tissue is present in the female. In this section the mammary tissue extending close to the popliteal lymph node of the hind limb. The epithelial components are embedded in white adipose tissue.

the underlying foot pad (Figure 10.12). The nail matrix is the proliferative zone which grows out continuously to form the nail plate.

10.3.2 *Mammary gland*

Mice show sexual dimorphism in mammary development from day 14 of gestation. Female mice have five sets of mammary glands and associated nipples whereas the male mouse has four mammary glands with no nipples and little or no remaining epithelial tissue in adulthood (Figure 10.13) (Kratochwil 1971).

In the nonpregnant mouse, the mammary tissue is composed of a branching ductular system embedded within adipose tissue (mammary fat pad).

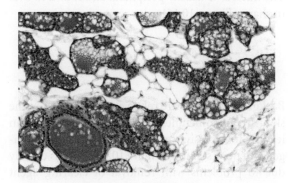

Figure 10.15   Lactating mammary gland.

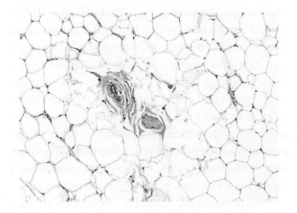

Figure 10.13   Small amount of residual mammary tissue may occasionally be sampled in adult male.

The ducts are lined by a simple cuboidal epithelium surrounded by a layer of myoepithelial cells, which are attached to a basal lamina (Figure 10.14). During pregnancy the epithelial component proliferates to form lobular alveoli. The alveoli have a single layer of epithelial cells, surrounded by a discontinuous layer of myoepithelial cells, which express cytokeratins and smooth muscle actin. The alveolar epithelial cells may contain clear lipid vacuole droplets and surround a central lumen containing milk secretion (Figure 10.15) (Oakes *et al.* 2006). Milk secretion starts on the day before parturition and continues for three weeks of lactation or until any pups are weaned or removed. The mammary gland starts to involute by apoptosis at weaning and returns to the pre-pregnant state within approximately two weeks (Richert *et al.* 2000).

10.3.3 *Adipose tissue*

White adipose tissue consists of large round to polygonal clear cells containing a single fat vacuole with a dense staining flattened nucleus displaced to the periphery. Scattered macrophages, dendritic cells and lymphocytes are a normal component of WAT (Lolmede *et al.* 2011) and lymph nodes are often found embedded in WAT (Figure 10.14). Other structures including blood vessels and ureters may be seen coursing between adipocytes depending on the site of sampling.

Brown adipose tissue consists of large polygonal cells containing multiple variably sized vacuoles. The nucleus is usually peripheral and round to oval in shape (Figure 10.16).

10.3.4 *Background lesions and artefacts*

Background lesions are relatively uncommon in the skin. It can sometimes be difficult to process skin leading to excessive separation of the dermal collagen and poor staining (Figure 10.17). This is usually due to problems with removal of clearing agent during process or difficulty in cutting cutaneous adipose tissue.

Some strains of mice, particularly from a C57Bl/6 background, are prone to developing alopecia and

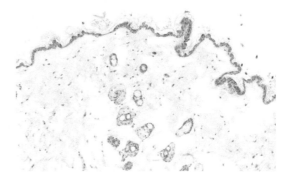

Figure 10.17   Poorly stained section of skin showing artefactual separation of collagen in dermis.

Figure 10.18   Example of ulcerative dermatitis from dorsal neck skin.

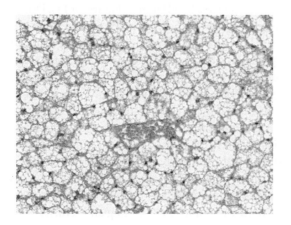

Figure 10.16   Interscapular brown adipose tissue with variable sized vacuoles.

ulcerative dermatitis, which often starts in the dorsal neck region (Sundberg *et al.* 2011) (Figure 10.18). Similar lesions may be seen with self-trauma related to ectoparasite infestations but these are increasingly rare in modern laboratory animal facilities. Traumatic lesions may occur due to accidental damage or as a result of fighting, particularly in some strains when they are group housed. These lesions may be confined to superficial skin damage, often to appendages, or may progress to pyogranulomas, often seen in the head region, or ascending urinary tract infections if the damage occurs in the genital area.

Skin tumours are relatively uncommon in mice and other differentials for skin masses should be considered including pyogranulomas, abscesses

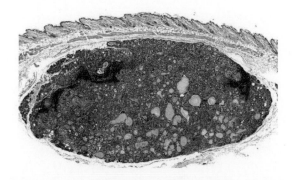

Figure 10.19 Skin masses can have a number of causes with mammary tumours being common in some strains of mice.

(of preputial gland for example) and mammary tumours, which are common in many mouse strains (Figure 10.19).

References

Cardiff, R.D., Anver, M.R., Gusterson, B.A. *et al.* (2000) The mammary pathology of genetically engineered mice: the consensus report and recommendations from the Annapolis meeting. *Oncogene* **19**, 968–988.

Casteilla, L., Pénicaud, L., Cousin, B. and Calise, D. (2008) Choosing an adipose tissue depot for sampling: factors in selection and depot specificity. *Methods in Molecular Biology* **456**, 23–38.

Dorfl, J. (1982) The musculature of the mystacial vibrissae of the white mouse. *Journal of Anatomy* **135**, 147–154.

Kratochwil, K.S. (1971) In vitro analysis of the hormonal basis for the sexual dimorphism in the embryonic development of the mouse mammary gland. *Embryology and Experimental Morphology* **25**, 141–153.

Lolmède, K., Duffaut, C., Zakaroff-Girard, A. and Bouloumié, A. (2011) Immune cells in adipose tissue: key players in metabolic disorders. *Diabetes and Metabolism* **4**, 283–290.

Majka, S.M., Barak, Y. and Klemm, D.J. (2011) Concise review: adipocyte origins: weighing the possibilities. *Stem Cells* **29** (7), 1034–1040.

Müller-Röver, S., Handjiski, B., van der Veen, C. *et al.* (2001) A comprehensive guide for the accurate classification of murine hair follicles in distinct hair cycle stages. *Journal of Investigative Dermatology* **117**, 3–15.

Oakes, S.R., Hilton, H.N. and Ormandy, C.J. (2006) Key stages in mammary gland development. The alveolar switch: coordinating the proliferative cues and cell fate decisions that drive the formation of lobuloalveoli from ductal epithelium. *Breast Cancer Research* **8**, 207–217.

Paus, R., Müller-Röver, S., Van Der Veen, C. *et al.* (1999) A Comprehensive Guide for recognition and classification of distinct stages of hair follicle morphogenesis. *Journal of Investigative Dermatology* **113**, 523–532.

Rasmussen, S.M., Young, L.J.T. and Smith, G.H. (2000) Preparing mammary gland whole mounts from mice, in *Methods in Mammary Gland Biology and Breast Cancer Research* (eds M.M. Ip and B.B. Asch), Kluwer Academic/Plenum Press, New York.

Richert, M.M., Schwertfeger, K.L., Ryder, J.W. and Anderson, S.M. (2000) An atlas of mouse mammary gland development. *Journal of Mammary Gland Biology and Neoplasia* **5**, 227–241.

Ruehl-Fehlert, C., Kittel, B., Morawietz. G. *et al.* (2003) Revised guides for organ sampling and trimming in rats and mice – part 1. *Experiments in Toxicologic Pathology* **55**, 91–106.

Sundberg, J.P. and King, L.E. (2000) Skin and its appendage: normal anatomy and pathology of spontaneous, transgenic and targeted mouse mutations, in *Pathology of Genetically Engineered Mice* (eds J.M. Ward, J.F. Mahler, R.R. Maronpot and J.P. Sundberg), Iowa State University Press, Ames, IA.

Sundberg, J.P., Peters, E.M.J. and Paus, R. (2005) Analysis of hair follicles in mutant laboratory mice. *Journal of Investigative Dermatology Symposium Proceedings* **10**, 264–270.

Sundberg, J.P., Taylor, D., Lorch, G. *et al.* (2011) Primary follicular dystrophy with scarring dermatitis in C57BL/6 mouse substrains resembles central centrifugal cicatricial alopecia in humans. *Veterinary Pathology* **48**, 513–524.

Waldén, T.B., Hansen, I.R., Timmons, J.A. *et al.* (2012) Recruited vs nonrecruited molecular signatures of brown, 'brite', and white adipose tissues. *American Journal of Physiology Endocrinology and Metabolism* **302** (1), E19–31.

Wong, P., Domergue, R. and Coulombe, P.A. (2005) Overcoming functional redundancy to elicit pachyonychia congenita-like nail lesions in transgenic mice. *Molecular and Cellular Biology* **25**, 197–205.

Chapter 11
The respiratory system

Elizabeth McInnes

Cerberus Sciences, Thebarton, Australia

11.1   Background and development

The respiratory system consists of the nasal cavity, larynx, trachea, major airways and lung parenchyma. The respiratory system is studied routinely in phenotyping studies and in more detail to test inhaled compounds for toxicity and for the evaluation of mouse models of human disease. For these reasons it is important to understand the anatomy and histology of the mouse respiratory tract. Nasal and pulmonary epithelia are able to metabolize xenobiotics and frequently demonstrate lesions from these compounds (Renne *et al.* 2009). Identification of the metabolic functions of the respiratory epithelia has resulted in great interest in the structures of the respiratory system.

11.2   Embryology

The mouse lung arises from the laryngotracheal groove. The embryonic phase of lung development begins with the formation of a groove in the ventral lower pharynx, the sulcus laryngotrachealis. The lung bud or true lung primordium emerges ventral to the caudal portion of the foregut. As the lung bud grows, its distal end enlarges to form the tracheal bud. The tracheal bud divides into two primary bronchial buds. The two main bronchi develop with the smaller bud on the left and the larger one on the right. At the same time the future trachea separates from the foregut through the formation of tracheoesophageal ridges, which fuse to form the tracheoesophageal septum.

In the mouse, three stages of development are recognized in the embryonic lung. The glandular period lasts until about day 10 and the lung at this stage consists of poorly defined connective tissue and proliferating columnar epithelium. During the cannalicular phase, which occurs from day 10 until a couple of days before birth, the bronchial tree and vascular systems start to develop. Alveoli start to form in the last couple of days before birth (alveolar period), but do not become fully differentiated until a few days postnatally. Postnatally (Figure 11.1) the alveolar spaces become defined by increased

A Practical Guide to the Histology of the Mouse, First Edition. Cheryl L. Scudamore.
© 2014 John Wiley & Sons, Ltd. Illustrations © Veterinary Path Illustrations, unless stated otherwise.
Published 2014 by John Wiley & Sons, Ltd. Companion Website: www.wiley.com/go/scudamore/mousehistology

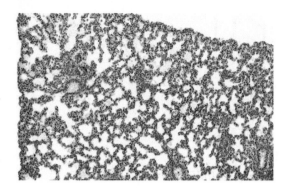

Figure 11.1 Section of lung from a mouse on postnatal Day 2. Note the relative thickness of the alveolar septae due to incomplete septation.

numbers of septae and the thickness of the septal walls decreases.

Considerable differences in lung volume and morphometric parameters exist between inbred mice. Adult C3H/HeJ mice (C3) have a 50% larger lung volume and 30% greater mean linear intercept than C57BL/6J (B6) mice (Soutiere and Mitzner 2006]). These differences are thought to occur because of lung air volume differences and different rates of alveolar septation (Soutiere and Mitzner 2006).

11.3 Anatomy and histology of the respiratory system

The respiratory system is divided into two major areas, the conducting portion and the respiratory portion (Gartner and Hiatt 1997). The conducting portion situated both outside and within the lungs, conveys air from the outside to the lungs. The respiratory portion, located within the lungs, functions in the exchange of oxygen for carbon dioxide and consists of the alveolar ducts and alveoli. The conducting portion of the respiratory system includes the nasal cavity, mouth, nasopharynx, pharynx, larynx, trachea, bronchi and bronchioles. The functions of the respiratory tract include exchange of air, control of acid/base balance of blood, excretion of substances and metabolism.

11.4 Upper respiratory tract

11.4.1 *Sampling the nasal cavity*

The cutting of standardised upper respiratory tract sections and the recognition of the normal histologic anatomy of cell populations within those sections is very important as it enables more accurate interpretation of the effects on cell types that will vary with location of the level of the section. The variations in the distribution of different epithelia within the nasal turbinates require consistent sectioning of the nasal turbinates and larynx. (Kittel., *et al.* 2004; Renne *et al.* 1992) (Figure 11.2a and b).

At necropsy, the lower jaw, calvarium and brain are removed and the nasal cavities should be perfused through the nasopharynx with formaldehyde. The entire head is then placed in fixative. Decalcification is performed in formic acid or Christensen's fixative and then tissues are removed from the decalcifying solution, processed and microtomed by standard procedures.

For diagnostic purposes, for instance to check for rhinitis, a single section through the nasal cavity may be made at the level of the upper incisor teeth.

If a complete evaluation of the nasal cavity is required then three or four standard sections of the mouse nasal turbinates may be cut with reference to the upper palate (Young 1981) and teeth (Renne *et al.* 2007) (Figure 11.2a). The first section (level 1) is cut at the level of the upper incisor teeth (Figure 11.3). The roots of the teeth are shown laterally. The paired vomeronasal organs (Jacobson's organ) and nasolacrimal ducts are present in level 1. A high volume of air passes through the central region of the nasal cavity so it is vulnerable to inhaled compounds and pathogens and is a common site for lesions (Renne *et al.* 2007).

The second section is taken at the level of the incisive papilla (Figure 11.4). At this level,

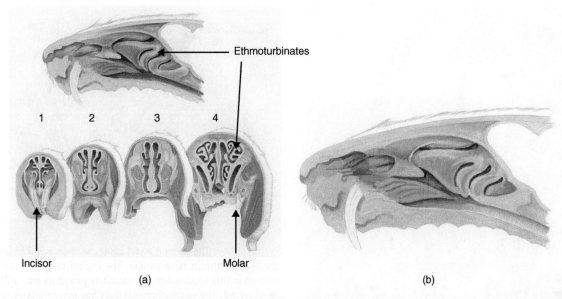

Figure 11.2 (a) Illustration of a longitudinal section of the nasal cavity and the four levels of sections, which may be taken to aid complete examination of the nasal cavity. (b) The distribution of epithelial types is illustrated progressing from squamous epithelium (orange) at the entrance through a narrow band of transitional epithelium (pink), to respiratory epithelium (lilac) and extensive areas of olfactory epithelium (blue) caudally.

bilateral communication with the oral cavity via the incisive ducts can be observed (Reznik 1990). The incisor teeth are still present in this section and the nasolacrimal duct may still be observed. The nasolacrimal duct has moved lateral to the incisor tooth. The olfactory epithelium in this level is especially susceptible to injury (Renne *et al.* 2007).

The third section is taken at the level of the second palatal ridge (Figure 11.5). This section includes the maxillary sinuses and Steno's glands. The glands surrounding the maxillary sinuses are known as Steno's glands. They are similar to serous salivary glands. The most frequent site of injury at level 3 in mice is the olfactory epithelium lining the dorsal medial meatus (Renne *et al.* 2007).

The fourth section is taken at the level of the first upper molar tooth (Figure 11.6). Here the nasopharyngeal duct is visible with the nasal-associated lymphoid tissue (NALT) present on either side of the nasopharyngeal duct. Bowman's glands produce mucus, are highly metabolic and are found in the lamina propria below the olfactory epithelium. Large and prominent nerve bundles in

the lamina propria can be observed as well as the olfactory bulbs of the brain.

11.4.2   *The nasal cavity*

The nasal turbinates, larynx, trachea and tracheal bifurcation and associated lymphoid tissue are often only examined histologically in inhalation toxicity studies. However consideration should be given to examining the nasal cavity in primary phenotyping screens, in cases of rhinitis secondary to ciliary defects or in accidental inhalation of bedding material or dusty food stuff. Inhalation of foreign material is (Figure 11.7) common and may be a cause of otherwise unexplained death in mice. Mice are obligate nose breathers and so obstruction to airflow due to rhinitis (or other space occupying lesion) can result in swallowing of air leading to a distended stomach and intestines which eventually compress the diaphragm causing death by asphyxiation.

The positions of the ethmoid turbinate, hard palate, incisor tooth, maxilloturbinate, naris and

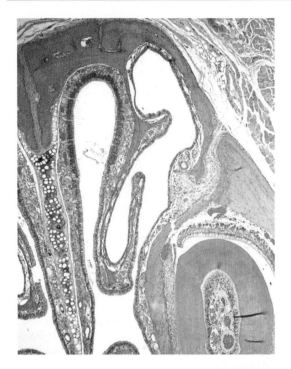

Figure 11.3 The first nasal cavity section (level 1) is cut at the level of the upper incisor teeth and includes the vomeronasal organ (not shown in this image). The most rostral location of level 1 is lined by stratified squamous epithelium in the ventral meatus. In level 1, transitional epithelium lines the lateral walls and lateral turbinate surfaces and olfactory epithelium can line the dorsal meatus. Respiratory epithelium lines the septum, dorsal meatus and medial aspects of the turbinates.

nasoturbinate nasal vestibule in the mouse are illustrated in Figure 11.2.

The mouse nasal cavity is divided into two main air passages by the nasal septum (Harkema *et al.* 2011). Each nasal passage extends from the nares to the nasopharynx caudally (Harkema *et al.* 2011). The nasal cavities contain a dorsal meatus, which becomes the dorsal ethmoid recess, a middle meatus that becomes the maxillary sinus and a ventral meatus, which terminates at the nasopharyngeal duct. The vestibule is located immediately after the nares before the main chambers of the nose (Harkema *et al.* 2011). Four epithelial types are found in the nasal cavity / nasal

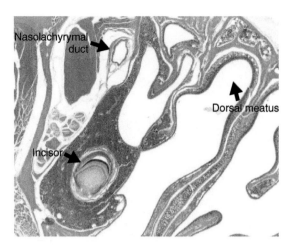

Figure 11.4 The second nasal cavity section is taken at the level of the incisive papilla. The incisor teeth are still present in this section and the nasolachrymal duct may still be observed. The nasolachrymal duct has moved lateral to the incisor tooth. The dorsal meatus is lined by olfactory epithelium. Squamous epithelium lines the nasopalatine duct ventrally. The respiratory epithelium lines the nasal septum, the turbinates and lateral walls and the transitional epithelium has disappeared at this level.

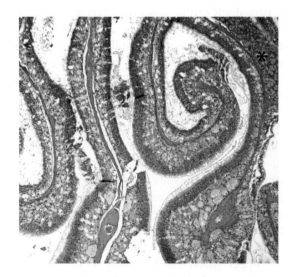

Figure 11.5 The third nasal cavity section is taken at the level of the second palatal ridge. This section includes the maxillary sinuses and Steno's glands. The majority of the ethmoid turbinates are lined by olfactory * epithelium. The maxillary sinuses are lined by respiratory epithelium.

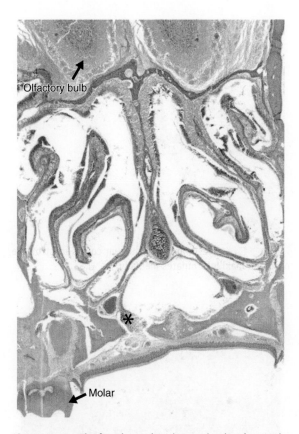

Figure 11.6   The fourth nasal cavity section is taken at the level of the first upper molar tooth. Here the nasopharyngeal duct is visible and is lined by respiratory epithelium. NALT is visible on ventrally either side of the nasopharyngeal duct. The epithelium in this region covering the ethmoid turbinates is almost entirely olfactory.

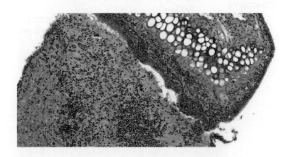

Figure 11.7   Accidental inhalation of bedding is common and causes rhinitis and this may be a cause of otherwise unexplained death in mice.

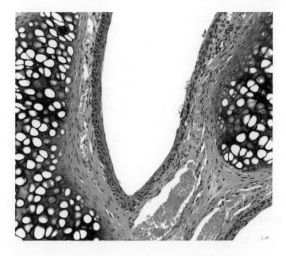

Figure 11.8   Stratified squamous epithelium lines the vestibule.

turbinates (Monticello *et al.* 1990). These include squamous epithelium, transitional epithelium, respiratory epithelium and olfactory epithelium. Stratified squamous epithelium lines the vestibule (Figure 11.8). It has a basal layer covered with several layers of squamous epithelial cells that become successively flatter towards the mucosal surface. The nasal vestibule requires squamous epithelium to protect the underlying tissues from injurious, inhaled substances. Squamous epithelium lines the ventral meatus, although at this level most of the turbinates are lined by respiratory epithelium.

Inhaled air moves from the vestibule to the nasal chambers, which are made up of nasal turbinates. The turbinates are bony structures that project into the lumen from the lateral walls and increase the surface area of the nose (Harkema *et al.* 2011). The squamous epithelium of the vestibule gives rise to transitional epithelium (Figure 11.9) before becoming respiratory epithelium, which lines the rostral turbinates. Thus, transitional epithelium lines the tips and lateral aspects of parts of the naso and maxilloturbinates and the lateral wall of the anterior nasal cavity. Transitional epithelium is one to two cells thick and is primarily made up of non-ciliated cuboidal and short columnar cells resting on basal cells, covered with microvilli. Transitional

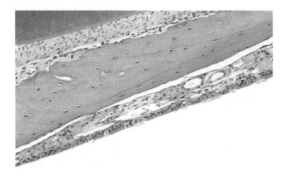

Figure 11.9   The squamous epithelium of the vestibule gives rise to transitional epithelium.

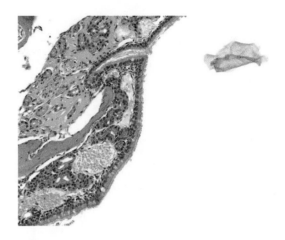

Figure 11.10   Respiratory epithelium covers most of the naso and maxilloturbinates medially.

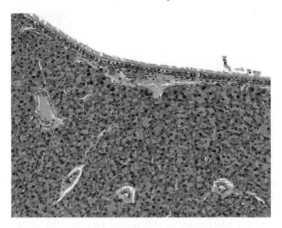

Figure 11.11   Steno's gland is situated in the lateral walls of the maxillary sinus.

epithelium contains few goblet cells (Harkema *et al.* 2011). The cells of the transitional epithelium contain abundant smooth endoplasmic reticulum (SER) in the cytoplasm and are an important store for xenobiotic-metabolizing enzymes such as cytochrome p450. The lamina propria below the squamous and transitional epithelia contains blood vessels, connective tissue, nerves and serous and mucous glands. Prominent 'swell bodies' (venous plexuses) are found in the maxilloturbinates and the lateral wall of the proximal nasal passage in mice (Harkema *et al.* 2011). These structures function to rehydrate the mucosa (Gartner and Hiatt 1997).

Respiratory epithelium (Figure 11.10) covers most of the naso and maxilloturbinates medially, the nasal septum and ventral ethmoid turbinates. Respiratory epithelium is composed largely of pseudostratified, ciliated and nonciliated cuboidal to columnar cells, goblet cells and basal epithelial cells. Solitary chemoreceptor and brush cells are also observed in respiratory epithelium (Harkema 2011). There is a gradual increase in goblet cells as the respiratory epithelium moves from the front of the nasal turbinates to the more posterior turbinates. The lamina propria below the respiratory epithelium contains serous and mucous glands. Steno's gland is situated in the lateral walls of the maxillary sinus (Figure 11.11).

Olfactory epithelium lines the ethmoid turbinates and some of the anterior dorsal meatus. Olfactory epithelium (Figure 11.12) is a pseudostratified

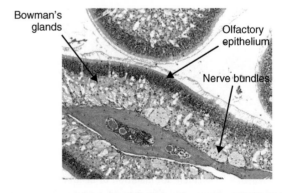

Figure 11.12   Olfactory epithelium is a pseudostratified columnar epithelium composed of tall sustentacular cells, olfactory neurons, and basal cells.

columnar epithelium. It is composed of tall sustentacular cells, olfactory neurons and basal cells. The sustentacular cells are columnar, secretory cells and support the olfactory neurons. In addition to supporting the olfactory neurons, the sustentacular cells also possess microvilli, which act together with the microvilli of the olfactory neurons. Large amounts of cytochrome p450 are found in the cytoplasm of the sustentacular cells (Harkema *et al.* 2011). The olfactory neurons are lined up between the sustentacular cells and their nuclei form a prominent layer, which is five to six nuclei thick. The olfactory sensory neurons are bipolar neuronal cells. The dendrites of the olfactory neurons extend above the epithelial surface and end in an olfactory knob which in turn, extends into the cilia onto the surface. The cilia and microvilli of the sensory neuron are involved in the function of smell. The axon of the olfactory sensory neuron extends through the basement membrane and joins other axons to form the nonmyelinated nerves bundles. The nonmyelinated nerve bundles penetrate the bony cribriform plate and extend into the olfactory bulb of the brain. The olfactory epithelium overlays a lamina propria made up of nerve bundles, blood vessels and tubulo-alveolar Bowman's glands. The Bowman's glands produce enzymes for the metabolism of xenobiotic compounds and secrete mucus that coats the surface epithelium (Renne *et al.* 2007). The paired vomeronasal glands (organ of Jacobson) (Figure 11.13) are situated in the base of the nasal septum. The vomeronasal organs are not present in humans. In animals, they are thought to play a role in recognition of pheromones and food flavour perception (Reznik 1990; Renne *et al.* 2009). The vomeronasal organs are lined by respiratory epithelium dorsal-medially and olfactory epithelium ventral-laterally. Neutrophils are found commonly in normal vomeronasal glands (Reznik 1990).

Important structures to evaluate histologically in the nasal cavities include the squamous epithelium, respiratory epithelium, transitional epithelium, olfactory epithelium, the vomeronasal organ, the nasolachrymal ducts, Bowman's glands in the olfactory epithelium (and other submucosal glands)

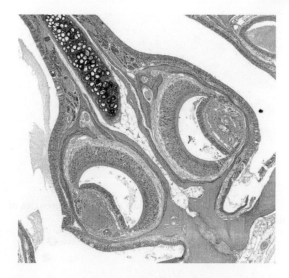

Figure 11.13 The paired vomeronasal glands (organ of Jacobson) are situated in the base of the nasal septum.

as well as the teeth and NALT (nasal associated lymphoid tissue). The areas of the rodent nasal mucosa that most frequently develop lesions in response to inhaled chemical compounds are the transitional and respiratory epithelia and the olfactory epithelium lining the dorsal medial meatus (Renne *et al.* 2007).

11.4.3 Lymphoid-associated tissue in the upper respiratory tract

Lymphoid tissues is generally present in the lamina propria adjacent to the nasopharyngeal duct of mice. Nasal associated lymphoid tissue in this area is similar to bronchus associated lymphoid tissue in the lungs. Recently, collection of NALT as a routine tissue in inhalation studies has been recommended (Renne *et al.* 2007). Lymphoepithelium covers the aggregates of nasal associated lymphoid tissue (NALT) (Figure 11.14), which are scattered throughout the lamina propria of the ventral aspects of the lateral walls at the entrance of the nasopharynx (Harkema 2011). The epithelium overlying the lymphoid aggregates is cuboidal. The position of NALT at the entrance of the nasopharynx allows the lymphoid tissue to be exposed to the majority of the inhaled foreign particles.

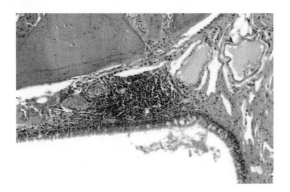

Figure 11.14 Lymphoepithelium covers the aggregates of nasal-associated lymphoid tissue (NALT).

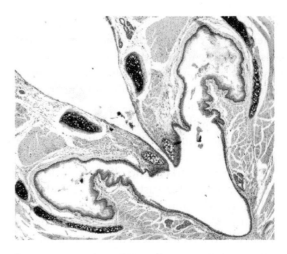

Figure 11.16 The larynx is a bilaterally symmetrical organ and is situated behind the base of the tongue and is ventral to the oesophagus.

11.4.4 The pharynx

The pharynx is made up of the nasopharynx (Figure 11.15), the oropharynx and the laryngopharynx and it extends from the mouth and nasal cavities to the larynx. The larynx serves as a common passage for food and air (Harkema, *et al.* 2011). The epithelia of the pharynx include squamous epithelium and respiratory epithelium. The Eustachian tube opens into the dorsal area of the nasopharynx – the oropharynx (Krinke 2004) – and seromucinous glands are present in the lamina propria of the nasopharynx. The nasopharynx begins at the choana and extends to the opening of the larynx and is lined by respiratory epithelium (Gartner and Hiatt 1997).

11.4.5 The larynx

The larynx (Figure 11.16) is a bilaterally symmetrical organ; it is situated behind the base of the tongue and is ventral to the oesophagus. The larynx is located between the pharynx and the trachea (Krinke 2004). The laryngeal wall is made up of epithelial lining, the cartilages with associated skeletal muscle and the outer connective tissue (Krinke 2004). The laryngeal cartilages include three paired cartilages (arytenoid, corniculate and cuneiform) and two single cartilages (thyroid and arytenoid) (Harkema *et al.* 2011). The epithelium changes from stratified squamous epithelium cranially to ciliated columnar respiratory epithelium caudally (Krinke 2004). Renne *et al.* (1992) describe methods for trimming and processing the larynx in order to obtain optimal histological sections (Renne *et al.* 1992). In specific inhalation studies in mice, six to eight coronal sections are often taken through the larynx, although some protocols section only through the base of the epiglottis to obtain the sensitive epithelium that overlies the submucosal glands. When a series of sections are taken, the coronal sections begin at the base of the epiglottis and continue through the thyroid and cricoid cartilages and arytenoid cartilages to the trachea.

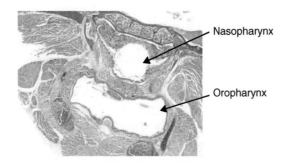

Nasopharynx

Oropharynx

Figure 11.15 The pharynx is made up of the nasopharynx, the oropharynx and the laryngopharynx and it extends from the mouth and nasal cavities to the larynx.

Specifically, the epithelial lining of a narrowly delineated region on the ventral floor of the larynges of rats and mice appears to be especially vulnerable to inhaled materials, and is recognized as a preferred site for histopathological evaluation in inhalation studies (Sagartz *et al.* 1992). This site is located at the base of the epiglottis, cranial to the ventral laryngeal diverticulum (ventral pouch) (Sagartz *et al.* 1992). The presence of underlying seromucinous glands is critical for histologic identification of this site (Sagartz *et al.* 1992).

The ventral glands at the base of the epiglottis possess excretory ducts, which exit onto the surface. The arytenoid cartilages are also a common site for toxicant-induced lesions. The floor / ventral surface of the larynx is covered by stratified, nonciliated epithelium with pseudostratified, ciliated columnar epithelium on either side. Mice have a ventral pouch situated between the caudal epiglottis and the trachea, above the thyroid cartilage, which is lined with ciliated, columnar and cuboidal cells (Harkema *et al.* 2011). The dorsal part of the lateral walls of the vestibule of the larynx are covered with stratified squamous epithelium (Reznik 1990). The vocal folds are covered with stratified squamous epithelium. A thorough knowledge of the normal laryngeal histology is important to detect subtle changes in epithelial type, which may be caused by xenobiotics (Lewis 1991; Renne *et al.* 1992). The squamous epithelium lining the rostral larynx is susceptible to injury because it is thin (Renne *et al.* 2007).

11.4.6   The trachea

The cellular composition of the respiratory, columnar, ciliated epithelium that covers the trachea (Figure 11.17) is essentially similar in all animal species (Reznik 1990). Clara cells are common in the trachea epithelium and function to produce surfactant (Suarez *et al.* 2012). Other cells in the tracheal epithelium include brush cells, neuroendocrine cells and basal cells (Suarez *et al.* 2012) and goblet cells are uncommon. Immune cells called globule leucocytes are scattered among respiratory epithelial cells lining the larynx and trachea of mice (Renne *et al.*

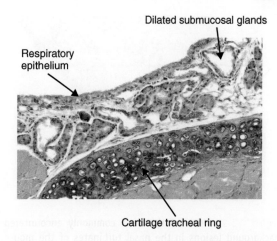

Figure 11.17   The trachea is covered with respiratory epithelium. Dilatation of submucosal glands is not uncommon.

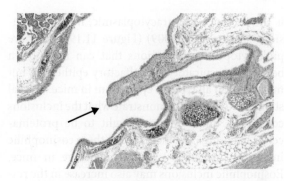

Figure 11.18   The respiratory epithelium lining the tip (arrow) of the tracheal bifurcation (carina) is a common site for lesions caused by irritant inhaled compounds.

2009). A few submucosal glands are noted in the lamina propria of the mouse. The tracheal bifurcation (carina) (Figure 11.18) is an impact area which is preferentially exposed to injurious xenobiotics and pathogens as the airflow moves down the trachea and into the bronchi.

11.4.7   Background lesions of the upper respiratory tract

One of the most commonly encountered background lesions in the nasal turbinates of the mouse

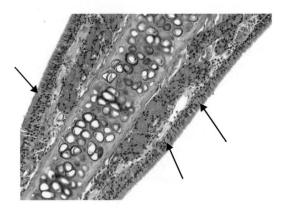

Figure 11.19 One of the most commonly encountered background lesions in the nasal turbinates of the mouse is the presence of intracytoplasmic hyaline inclusions (arrows).

is the presence of intracytoplasmic, hyaline inclusions (Renne *et al.* 2009) (Figure 11.19). They are pink cytoplasmic inclusions that can most often be seen in respiratory and olfactory epithelium but may occur in all types of epithelium in mice. Special stains have failed to demonstrate what the inclusions consist of, but they are thought to be proteinaceous in nature. The incidence of these eosinophilic inclusions gradually increases with age in mice. Eosinophilic inclusions may also increase in the respiratory and olfactory epithelium (as well as in the associated seromucinous glands in the lamina propria) due to mild and chronic irritation over a period of time. Ward *et al.* (2001) have demonstrated that the nasal epithelial inclusions react with antibodies to the Ym1 sequence of the protein Ym2, a member of the chitinase family.

Another common change in the nasal turbinates of mice, which increases in severity with age, is the accumulation of hyaline pink material in the nasal ventral septum (Herbert and Leininger 1999). The material appears similar to amyloid but does not stain positive with common special stains for amyloid. Occasional concretions of mineral (corpora amylacea) are often noted within or immediately below the epithelia. These are considered to be normal unless greatly increased in number. Dilated submucosal glands in the larynx and trachea are observed commonly in mice (Lewis 1991) (Figure 11.17). Inflammatory changes in the turbinates and larynx occasionally occur in association with foreign bodies, infection or injury (Herbert and Leininger 1999). Foreign material is seen in the lumen of the ventral pouch of the larynx, associated with an inflammatory reaction and hyperplasia of the laryngeal epithelium (Figure 11.20).

Lymphocytic or eosinophilic infiltrates into the nasal cavity – i.e. below the squamous, transitional, respiratory, olfactory and glandular epithelium of the nasal cavity is common and may indicate inflammation or lymphoid hyperplasia. Eosinophilic infiltrates may be caused by parasites or allergic reactions (Renne *et al.* 2009).

The presence of clusters of neuroendocrine cells within the respiratory or olfactory epithelium of the nasal cavity and within the lungs may be observed (Haworth *et al.* 2007).

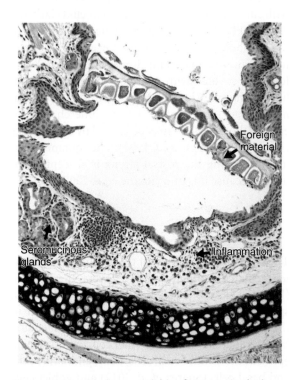

Figure 11.20 Foreign material is often seen in the lumen of the ventral pouch of the larynx, associated with an inflammatory reaction. Seromucinous glands help identify this vulnerable region of the larynx.

11.5   Lower respiratory tract

11.5.1   *Sampling techniques and morphometry of the lower respiratory tract*

Tracheal instillation of the lungs with buffered formalin at necropsy is an accepted technique to improve the histology of the pulmonary architecture (Hausmann *et al.* 2004) and should be used for all studies in mice. Intratracheal instillation of 10% nonbuffered formalin (NBF) (Braber *et al.* 2010) is recommended for excellent lung histopathological detail. Braber *et al.* (2010) have demonstrated that intratracheal instillation of 10% NBF and paraffin embedding is superior to plastic embedding and Carnoy's instillation via tracheal instillation or fixed volume fixation. The tracheal instillation may be performed after removing the lungs from the thoracic cavity or *in situ* (which prevents the collapse of the lungs, which occurs when the thoracic cavity is opened and the negative pressure is removed) (Braber *et al.* 2010). Failure to inflate lung tissue at necropsy results in collapse of the interstitial, alveolar tissue and the lung appears hypercellular mimicking pneumonic changes (Renne *et al.* 2009) (Figure 11.21). However, it is important to recognize that instillation of fixative may result in a lesion that resembles perivascular oedema and can also disperse exudates and cellular infiltrates. Formalin-instilled lung tissue is characterized by flattened alveolar walls, displacement of alveolar macrophages (Matulionis 1986) and inflammatory cells present within the alveolar spaces in the absence of fluid exudate (Turner *et al.* 1990). Although instillation of fixative via the trachea is adequate for most routine examinations it may sometimes be necessary to avoid the creation of artefacts. One method to avoid artefacts is to inflate the lung with air to a volume equal to the total lung capacity. The lung is then placed in phosphate buffered saline and the lung tissue is fixed for 4 min at 60 °C in a microwave oven (Turner *et al.* 1990). Alternatively the lungs can be fixed by vascular perfusion with fixative (Brain *et al.* 1984), which can reduce the disruption to alveolar macrophages, cellular infiltrates and exudates. Lung tissue must be handled with care at necropsy to avoid collapse and damage by forceps resulting in 'pinch' defects which do not resolve during processing (Thompson and Luna 1978).

Separate mouse lung lobes may be harvested and processed for histopathological analysis or the whole lung may be harvested and processed. Analysis of the entire trachea and lung is advantageous if the research model involves quantitative analysis of tumour development (Lai *et al.* 1993). Mouse models of human diseases such as asthma or the testing of new inhalation drugs may benefit from consistent sampling of only a proportion of mouse lung lobes as described by Kittel *et al.* (2004).

Although H&E staining is sufficient for routine histological examination, a number of special stains may be helpful for investigating commonly induced changes such as goblet cell hyperplasia, smooth muscle hyperplasia and fibrosis. Alcian blue / periodic acid Schiff's stain (AB/PAS) is useful for identification of mucin in goblet cells in respiratory epithelia (Tomlinson *et al.* 2010) and scoring systems have been suggested to quantify this (Ahn *et al.* 2007). Masson's Trichome (Figure 11.22) and picro-sirius red stains have been recommended for detection of collagen in fibrosis (Gregory *et al.* 2010; Tomlinson *et al.* 2010). Smooth muscle detection in lung tissue

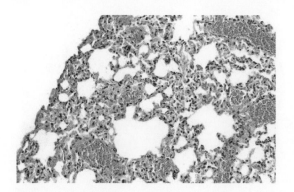

Figure 11.21   Failure to perfuse lung tissue at necropsy results in collapse of the interstitial, alveolar tissue and the histological appearance may appear to mimic pneumonia.

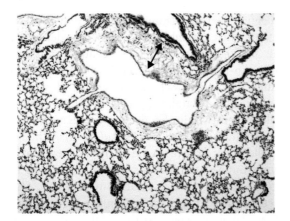

Figure 11.22 Massons trichrome stain can be used to demonstrate collagen (green). This image also shows artefactual separation of tissue around a blood vessel (double headed arrow) which needs to be differentiated from perivascular oedema.

may be performed using immunohistochemistry with an anti- alpha smooth muscle actin (α-SMA) (Johnson *et al.* 2004; Fattouh *et al.* 2011).

There are many papers describing morphometric analysis of mouse airways and pathological associated changes within the lungs (for example Ellis *et al.* (2003)). All techniques require standardised lung inflation and sampling as described. Histology of the lung offers higher resolution and the qualitative differentiation of tissues by staining but there is a loss of 3D tissue relationships and techniques such as microcomputed tomography, if available, may allow for more detailed integration of morphometric data with the spatial complexity of lung structure (Vasilescu *et al.* 2012).

11.5.2 *Lung*

The lungs are present in the thoracic cavity and are covered by the visceral pleura (Krinke 2004). The functions of the lung include gaseous exchange and neuroendocrine functions. The left lung lobe in the mouse forms a single lobe whereas the right lobe is divided up into the cranial, middle, caudal and accessory lobes (Krinke 2004). The conducting tract of the lower respiratory system consists of the trachea, which gives rise to the two main bronchi at the tracheal bifurcation. The term 'bronchus' refers to airways below the trachea, the walls of which contain cartilage, smooth muscle and submucosal glands (Renne *et al.* 2009). The bronchiolar mucosa is lined with columnar, ciliated epithelium and Clara cells and submucosal glands are not present (Renne *et al.* 2009). The bronchi divide to form secondary bronchi, then bronchioles and terminal bronchioles and alveolar ducts. The extrapulmonary bronchi in the mouse have cartilage rings (Figure 11.23) and intrapulmonary bronchi in mice do not have cartilagenous rings. Mice do not have distinct respiratory bronchioles (Suarez *et al.* 2012). In the mouse lung, the terminal bronchiole leads directly into the alveolar duct (Figure 11.24) with no recognizable respiratory bronchioles.

The respiratory portion of the respiratory tract consists of the alveolar ducts and alveoli. The alveolar ducts are lined by type-1 pneumocytes, with alveoli opening directly off from the walls. The epithelium of the mouse bronchi changes from columnar to cuboidal as one moves distally down the conducting tree with increasing numbers of Clara cells present. The large airways are lined by columnar epithelium, mainly nonciliated Clara cells (Figure 11.25), ciliated cells, neuroendocrine cells, mucous cells and brush cells (Krinke 2004). Clara cells can be visualized using immunohistochemistry (IHC) and function in innate defence and epithelial

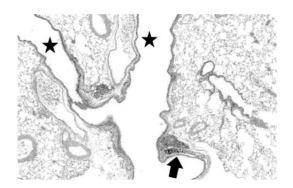

Figure 11.23 Extrapulomary bronchi in the mouse have cartilage rings (arrow) which are not present in the intrapulmonary bronchi (stars).

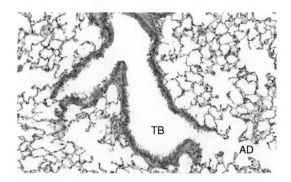

Figure 11.24 In the mouse lung, the terminal bronchiole (TB) leads into the alveolar duct (AD). The terminal bronchiole is predominantly lined by non ciliated Clara cells which have distinctive apical cytoplasmic blebs.

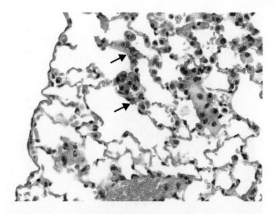

Figure 11.26 The alveolar epithelium is made up of flattened squamous pulmonary type-1 cells and cuboidal type-2 pneumocytes (arrows). Type 2 pneumocytes may be hard to identify in normal lung but become prominent when they undergo hyperplasia.

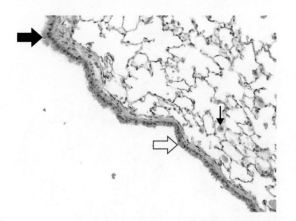

Figure 11.25 Distally in the bronchiolar system the ciliated epithelium (thick black arrow) is replaced by predominantly non-ciliated Clara cells (thick white arrow). Occasional macrophages (thin arrow) can normally be seen within alveoli.

repair (Kelly *et al.* 2012). Alveolar pores called the pores of Kohn link the alveoli and maintain an equal pressure between linked alveoli and prevent collapse (Suarez *et al.* 2012) – these cannot be visualized with standard light microscopic techniques.

The alveolus is the functional structure of the lung. The alveolar epithelium is made up of flattened squamous pulmonary type-1 cells and cuboidal type-2 pneumocytes (Figure 11.26). The type-1 pneumocytes are involved in gaseous exchange whereas the type-2 pneumocytes produce surfactant. Oxygen and carbon dioxide must cross the alveolar wall during gas exchange. Surfactant is composed of phospholipids, which reduce surface tension when the alveolus deflates and prevents alveolar collapse. The bronchoalveolar junction is important in inhalation toxicity studies as this is the first area that the inhaled air and compound will encounter as it travels down the respiratory tract. The inhaled air in the alveolus has close proximity to the alveolar surface epithelium in particular with the red blood cells of the small alveolar blood vessels. Alveolar macrophages lie free in the alveolar space (Figure 11.25). Some alveolar macrophages may be located within the connective tissue of all respiratory passageways. Alveolar macrophages are responsible for engulfing foreign material within the lung. Myofibroblasts within the alveolar walls control the alveolar volume. The lymphoid tissue in the lung is called bronchus-associated lymphoid tissue (BALT) and is barely visible in the healthy mouse lung. The BALT becomes prominent in respiratory infections, particularly viral infections (Figure 11.27).

11.5.3 Background lesions in the lung

Small foci made up of osteoid or bone with mineralization or calcification are common in the

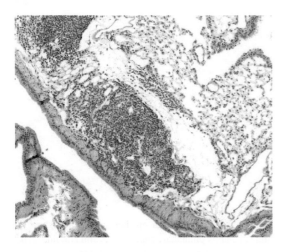

Figure 11.27   The BALT becomes prominent in respiratory infections, particularly viral infections.

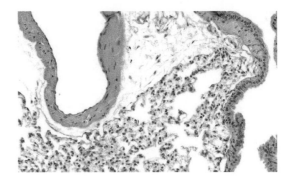

Figure 11.29   The walls of the major pulmonary veins of the lungs in mice contain cardiac muscle. Note this animal was perfused via the heart giving the alveolar septae an empty appearance.

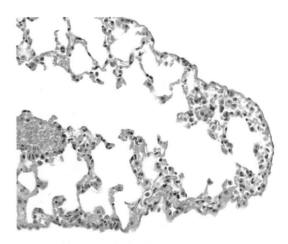

Figure 11.30   Focal accumulations of macrophages (alveolar histiocytosis) are a common incidental finding in the lungs and can be observed in subpleural areas of ageing mouse lungs.

mouse lung (Figure 11.28). There is usually no reaction in the surrounding lung tissue. The walls of the major pulmonary veins of the lungs in mice contain cardiac muscle (Percy and Barthold 2007) (Figure 11.29). Focal accumulations of macrophages (alveolar histiocytosis) (Figure 11.30) are a common incidental finding in the lungs (Renne *et al.* 2009) and can be observed in subpleural areas of

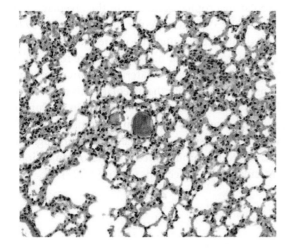

Figure 11.28   Small foci made up of osteoid or bone with mineralization or calcification are common in the mouse lung.

ageing mice lungs. Laryngeal and tracheal cartilage displays age-related mineralization and occasionally ossification. Lymphoid aggregates are commonly found in the mediastinal tissue of normal mice (Figure 11.31). Agonal haemorrhage may be related to the method of euthanasia and can be distinguished from pathological haemorrhage by the absence of any other pathology including inflammation and the absence of pigmented macrophages, which would indicate chronicity (Figure 11.32).

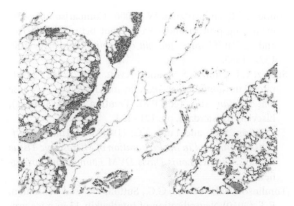

Figure 11.31   Lymphoid aggregates are commonly seen in the mediastinal tissue and around the pleura.

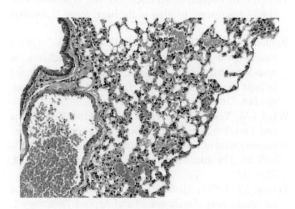

Figure 11.32   Agonal haemorrhage and congestion is common in mouse lungs and may be related to the method of euthanasia. It can be distinguished from pathological haemorrhage by the absence of pigmented macrophages or inflammation.

References

Ahn, J.H., Kim, C.H., Kim, Y.H. *et al.* (2007) Inflammatory and remodeling events in asthma with chronic exposure to house dust mites: a murine model. *Journal of Korean Medical Science* **22**, 1026–1033.

Brain, J.D., Gehr, P., Kavet, R.I. (1984) Airway macrophages. The importance of the fixation method. *American Review of Respiratory Disorders* **129**, 823–826.

Braber, S., Verheijden, K.A., Henricks, P.A. (2010) A comparison of fixation methods on lung morphology in a murine model of emphysema. *American Journal of Physiology Lung Cellular and Molecular Physiology* **299**, L843–851.

Ellis, R., Leigh, R., Southam, D. *et al.* (2003) Morphometric analysis of mouse airways after chronic allergen challenge. *Laboratory Investigation* **83**, 1285–1291.

Fattouh, R., Al-Garawi, A., Fattouh, M. *et al.* (2011) Eosinophils are dispensable for allergic remodeling and immunity in a model of house dust mite-induced airway disease. *American Journal of Respiratory and Critical Care Medicine* **183**, 179–188.

Gartner, L.P. and Hiatt, J.L. (1997) Respiratory system, in *Color Textbook of Histology* (eds L.P. Gartner and J.L. Hiatt) W.B. Saunders, Philadelphia, PA, pp. 284–302.

Gopinath, C., Prentice, D.E. and Lewis, D.G. (1987) The respiratory system, in *Atlas of Experimental Toxicological Pathology*, vol. **13** (ed. G.A. Gresham), MTP Press, Lancaster, pp. 22–42.

Gregory, L.G., Mathie, S.A., Walker, S.A. *et al.* (2010) Overexpression of Smad2 drives house dust mite-mediated airway remodeling and airway hyperresponsiveness via activin and IL-25. *American Journal of Respiratory and Critical Care Medicine* **182**, 143–154.

Harkema, J.R., Carey, S.A., Wagner, J.G. *et al.* (2011) Nose, simus, pharynx and larynx, in *Comparative Anatomy and Histology: A Mouse and Human Atlas* (eds P.M. Treuting and S. Dintzis), Elsevier, Amsterdam, pp. 71–94.

Hausmann, R., Bock, H., Biermann, T. *et al.* (2004) Influence of lung fixation technique on the state of alveolar expansion – a histomorphometrical study. *Legal Medicine* (Tokyo) **6**, 61–65.

Haworth, R., Woodfine, J., McCawley, S. *et al.* (2007) Pulmonary neuroendocrine cell hyperplasia: identification, diagnostic criteria and incidence in untreated ageing rats of different strains. *Toxicologic Pathology* **35**, 735–740.

Herbert, R.A. and Leininger, J.R. (1999) Nose, larynx, and trachea, in *Pathology of the Mouse, Reference and Atlas* (R.R. Maronpot, G.A. Boorman, B.W. Gaul), Cache River Press, Vienna, pp. 259–292.

Johnson, J.R., Wiley, R.E., Fattouh, R. *et al.* (2004) Continuous exposure to house dust mite elicits chronic airway inflammation and structural remodeling. *American Journal of Respiratory and Critical Care Medicine* **169**, 378–385.

Kelly, F.L., Kennedy, V.E., Jain, R. *et al.* (2012) Epithelial Clara cell injury occurs in bronchiolitis obliterans syndrome after human lung transplantation. *American Journal of Transplantation* **12**, 3076–3084.

Kittel, B., Ruehl-Fehlert, C., Morawietz, G. *et al.* (2004) Revised guides for organ sampling and trimming in rats and mice – part 2. A joint publication of the RITA and NACAD groups. *Expimental Toxicologic Pathology* **55**, 413–431.

Krinke, G.J. (2004) Normative histology, in *The Laboratory Mouse* (eds H. Hedrich and G. Bullock), Elsevier, New York, pp. 146–148.

Lai, W.C., Linton, G., Bennett, M. and Pakes, S.P. (1993) Genetic control of resistance to Mycoplasma pulmonis infection in mice. *Infection and Immunity* **61**, 4615–4621.

Lewis, D.J. (1991) Morphological assessment of pathological changes within the rat larynx. *Toxicologic Pathology* **19**, 352–357.

Matulionis, D.H. (1986) Lung deformation and macrophage displacement in smoke-exposed and normal mice (*Mus musculus*) following different fixation procedures. *Virchow's Archive A: Pathological Anatomy and Histopathology* **410**, 49–56.

Monticello, T.M., Morgan, K.T. and Uraih, L. (1990) Non-neoplastic nasal lesions in rats and mice. *Environmental Health Perspectives* **85**, 249–274.

Percy, D.H. and Barthold, S.W. (2007) *Pathology of Laboratory Rodents and Rabbits*, 3rd edn, Wiley-Blackwell.

Renne, R., Brix, A., Harkema, J. *et al.* (2009) Proliferative and nonproliferative lesions of the rat and mouse respiratory tract. *Toxicologic Pathology* **37**, 5S–73S.

Renne, R.A., Gideon, K.M., Harbo, S.J. *et al.* (2007) Upper respiratory tract lesions in inhalation toxicology. *Toxicologic Pathology* **35**, 163–169.

Renne, R.A., Gideon, K.M., Miller, R.A. *et al.* (1992) Histologic methods and interspecies variations in the laryngeal histology of F344/N rats and B6C3F1 mice. *Toxicologic Pathology* **20**, 44–51.

Reznik, G.K. (1990) Comparative anatomy, physiology, and function of the upper respiratory tract. *Environmental Health Perspectives* **85**, 171–176.

Sagartz, J.W., Madarasz, A.J., Forsell, M.A. *et al.* (1992) Histological sectioning of the rodent larynx for inhalation toxicity testing. *Toxicologic Pathology* **20** (1), 118–121.

Soutiere, S.E. and Mitzner, W. (2006) Comparison of postnatal lung growth and development between C3H/HeJ and C57BL/6J mice. *Journal of Applied Physiology* **100**, 1577–1583.

Suarez, C.J., Dintzis, S.M. and Frevert, C.W. (2012) Respiratory, in *Comparative Anatomy and Histology: A Mouse and Human Atlas* (eds P.M. Treuting and S. Dintzis), Elsevier, Amsterdam, pp. 121–134.

Thompson, S.W. and Luna L.G. (1978) *An Atlas of Artifacts Encountered in the Preparation of Microscopic Tissue Sections, Charles Louis Davis DVM Foundation*, Springfield, IL.

Tomlinson, K.L., Davies, G.C., Sutton, D.J. and Palframan, R.T. (2010) Neutralisation of interleukin-13 in mice prevents airway pathology caused by chronic exposure to house dust mite. *PLoS One* **5**, 1–7.

Turner, C.R., Zuczek, S., Knudsen, D.J. and Wheeldon, E.B. (1990) Microwave fixation of the lung. *Stain Technology* **65**, 95–101.

Vasilescu, D.M., Klinge, C., Knudsen, L. *et al.* (2012) Stereological assessment of mouse lung parenchyma via non-destructive multi-scale micro CT imaging validated by light microscopic histology. *Journal of Applied Physiology* **114**, 716–724.

Ward, J.M., Yoon, M., Anver, M.R. *et al.* (2001) Hyalinosis and Ym1/Ym2 gene expression in the stomach and respiratory tract of 129S4/SvJae and wild-type and CYP1A2-null B6, 129 mice. *American Journal of Pathology* **158**, 323–332.

Young, J.T. (1981) Histopathological examination of the rat nasal cavity. *Fundamental and Applied Toxicology* **1**, 309–312.

Chapter 12
Special senses

Cheryl L. Scudamore
Mary Lyon Centre, MRC Harwell, UK

12.1 Background and development

The embryonic development of the eye is similar in all species. The retina and pigment layers in the eye derive from neurectoderm, the lens and cornea from surface ectoderm and the uvea and sclera from mesoderm. Initially the optic vesicle forms from bilateral outpouchings of forebrain neurectoderm. The optic vesicle expands retaining a connective stalk to the developing brain, which will become the optic nerve. The surface ectoderm overlying the optic vesicle thickens to form the lens placode, which subsequently invaginates into the optic vesicle. The surface ectoderm differentiates into two regions which will ultimately form the lens (lens vesicle) and the cornea and eyelids. The surrounding mesoderm condenses to form the outer uveal and sclera coats. Detailed descriptions and timing of these changes can be found in Kaufmann and Bard (1999).

The mouse eye is not fully developed at birth and a detailed description of the changes that occur is beyond the scope of this chapter but is reported by Smith *et al.* 2002a. Briefly, the mouse is born with closed eyelids, which open at day 12–14 postnatally. All structures in the eye show immature features including increased stromal density, presence of hyaloid vasculature, compressed appearance of retinal layers and rapid lens growth including mitotic activity) (Figure 12.1). As development of the eye is largely complete by 3 weeks of age, with the exception of retained remnants of the hyaloid blood vessels, retention of immature features after this time may be considered unusual.

12.2 Sampling technique for the eye

There are a number of approaches to sampling the eye and associated intraorbital glands (intraorbital lacrimal gland and Harderian gland). The head may be fixed whole and the eyes left in the head and sectioned with a whole head cross-section following

---

A Practical Guide to the Histology of the Mouse, First Edition. Cheryl L. Scudamore.
© 2014 John Wiley & Sons, Ltd. Illustrations © Veterinary Path Illustrations, unless stated otherwise.
Published 2014 by John Wiley & Sons, Ltd. Companion Website: www.wiley.com/go/scudamore/mousehistology

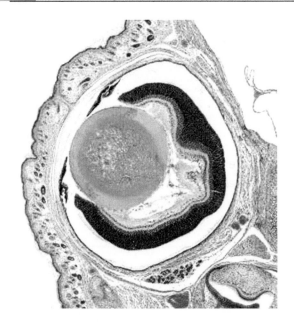

Figure 12.1　Early postnatal eye with dense retinal layers and remnants of hyaloid vessels.

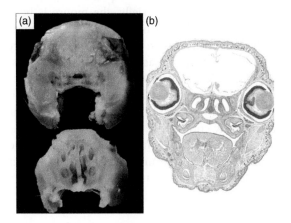

Figure 12.2　(a) and (b) Cross section through head and eye. Microscope image of early postnatal mouse.

decalcification (Figure 12.2a) or the eyes may be removed by enucleation (Smith *et al.* 2002a). Sectioning the eyes in the head has the virtue of being technically easier at necropsy, although it may not be possible to get a complete section through the globe and optic nerve and in addition the eye may not be well fixed if standard formalin fixative is used. An alternative approach is to bisect the head longitudinally after decalcification, embed

on the cut surface so that longitudinal sections can be made through the eye (Ramos *et al.* 2011) (Figure 12.3). The eye may also be dissected out fresh at the time of necropsy by careful enucleation (Smith *et al.* 2002b; Escher and Schorderet 2011) or after fixation of the head. It is important that if the eye is dissected fresh at the time of necropsy that excessive traction is not placed on the optic nerve as this may cause extensive artefact. The eye can be fixed at this stage by immersion in an appropriate fixative or further dissection can be undertaken if specific tissues are required for additional analysis, for example RNA extraction (Escher and Schorderet 2011). While fixation in Davidson's fixative (Latendresse *et al.* 2002) is advocated to

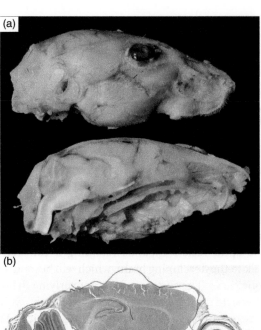

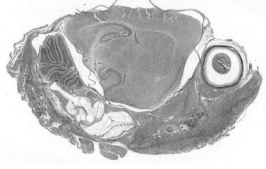

Figure 12.3　(a) and (b) Longitudinal section through head can be used to sample inner ear and eye. Sections without artefact may be hard to achieve in animals older than about 14 days post natal.

optimize retinal morphology, in practice the small size of the mouse eye means that adequate fixation, at least for primary morphological phenotyping screens, may be achieved using routine neutral buffered formalin or paraformaldehyde (Dubielzig *et al.* 2010, Ch. 1). There is no need to inject formalin into the globe or open the eye to allow penetration of fixation.

If dissected carefully, a length of optic nerve may be left attached to the globe and can be sectioned along with the rest of the eye if carefully embedded. Alternatively the optic nerve may be dissected at the time of necropsy and embedded separately (Smith *et al.* 2002b). Similarly the Harderian gland may be dissected separately or blocked in association with the globe.

Ideally the eye should be sectioned in a vertical longitudinal plane (Moraweitz *et al.* 2004) through the optic nerve. If precise orientation of the eye is required then it is advisable to mark the temporal, nasal, dorsal or ventral points of the globe at the time of dissection using a suitable indelible dye or tattoo ink (Figure 12.4). The distribution of cone photoreceptor cells varies between the dorsal and ventral aspects of the mouse retina (Szel *et al.* 1992) and so it may be necessary to orientate the eye when embedding to ensure that both aspects of the retina are examined (Figure 12.5). In general, haematoxylin and eosin stain is sufficient for examination of eye histology. However periodic acid – Schiff (PAS)/Alcian Blue may be useful to demonstrate basement membranes, trichrome stains to show connective tissue and Luxol fast blue/cresyl violet stain may be useful to demonstrate myelinated fibres in the optic nerve (Dubielzig *et al.* 2010, Ch. 1) (Figure 12.6). If necessary, techniques can be used to bleach out melanin in the retinal pigment epithelium (RPE) to allow cellular morphology to be examined in more detail.

12.3   Artefacts

The eye is prone to artefacts, which can occur due to handling of the eye in life, during dissection or during fixation and processing.

Figure 12.4 Eye can be marked with indelible ink by gently protruding the eye by applying pressure to draw back the eyelids. The eye should be gently dried by dabbing with tissue before repeatedly applying indelible ink to produce a clear spot.

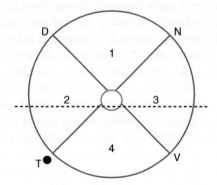

Figure 12.5 Diagram to show plane of section (dotted line) to ensure all quadrants of retina examined from the right eye marked as shown in Figure 12.4. To embed the right eye the tissue should be placed in the embedding mould with the optic nerve (open circle) to the left and the dot ● facing directly away from the histologist and held with the optic nerve and dot parallel to the mould base. The eye should be rolled down so that the dot is at 45° to the horizontal and held in this position until the wax has cooled. The paraffin block obtained should be trimmed until the optic nerve is visible and sections taken from this level. The same principle can be applied to the left eye if required. D = dorsal, V = ventral, N = nasal, T = temporal.

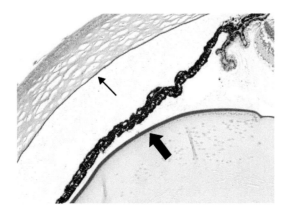
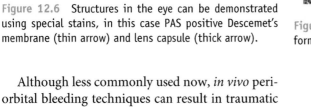

Figure 12.6   Structures in the eye can be demonstrated using special stains, in this case PAS positive Descemet's membrane (thin arrow) and lens capsule (thick arrow).

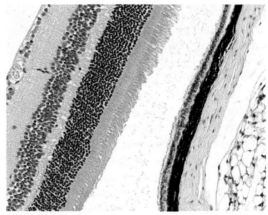

Figure 12.7   Artefactual detached retina is common in formalin fixed tissue.

Although less commonly used now, *in vivo* peri-orbital bleeding techniques can result in traumatic damage to the structures associated with the eye leading to inflammation of the optic nerve, surrounding muscles and Harderian gland (Van Herck *et al.* 1992; Williams 2002; Heimann *et al.* 2009).

The eye is easily damaged during dissection and artefacts related to handling are common and may include obvious traumatic lesions and others that may be more subtle, for example, detachment of the retina. An artefactually detached retina can be distinguished from a pathological condition due to the absence of proteinaceous fluid, attachment of receptor process fragments to the retinal pigment epithelium (RPE) and lack of RPE hypertrophy (Figure 12.7). Detached retina and poor retinal morphology are also common unwanted features of formalin fixation. Crushing of the optic nerve during dissection can lead to artefactual displacement of myelin between the choroid and retina in the region adjacent to the nerve.

The eye is prone to multiple artefacts following fixation and processing for histology. Separation of the collagen lamellae in the cornea is commonly seen (Figure 12.8) and can be particularly marked in eyes stored in modified Davidson's fixative (Latendresse *et al.* 2002). The lens is a relatively hard tissue compared to the other structures of the eye, which makes it difficult to section. As a result of this and

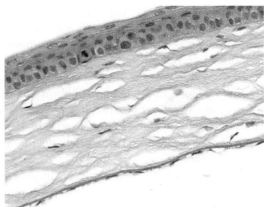

Figure 12.8   Separation of collagen lamellae in cornea of Davidson's fixed eye. Also showing mitotic figure in basal layer of corneal epithelium.

fixation, a range of artefacts including separation of the capsule, cracking and cavitation of the body of the lens and the appearance of granular material can occur, which all need to be distinguished from the cataract formation that is common in ageing mice (Figure 12.9). Retinal folds and rosettes are common background changes but may also be caused by rough handling during processing (Figure 12.10). In addition without careful orientation at embedding oblique sections can lead to apparent thickening or thinning of various structures (Smith *et al.* 2002b).

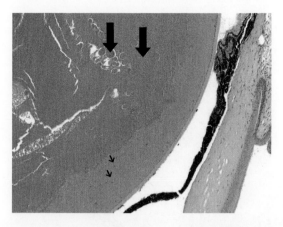

Figure 12.9 Cataracts can be confirmed by features such as swollen lens fibres (thick arrows) and disorganized nuclei (thin arrows).

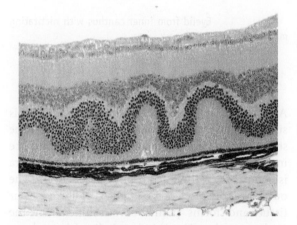

Figure 12.10 Rosettes and folds in the retina may be seen as incidental findings and can be due to inappropriate handling of tissue.

12.4   Anatomy and histology of the eye and associated glands

The eye is a complex structure, which is contained within a tough connective tissue capsule called the sclera. The integrity of the sclera is interrupted cranially by the transparent cornea and caudally by the optic nerve. The sclera is joined at the limbus by the palpebral conjunctiva of the eyelid (Figure 12.11).

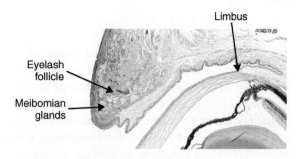

Figure 12.11   Eyelid.

The sclera is lined by the uvea, which consists of the vascular elements of the eye – the iris, ciliary body and choroid – which is in turn lined by the retina. The lens is suspended between zonules linked to the ciliary body.

The inner eye can also be divided anatomically into three major compartments. The most cranial (anterior) is the anterior compartment, which lies between the inner surface of the cornea and the cranial surface of the iris. The posterior compartment lies between the iris and the lens. The vitreous compartment which lies between the lens and retina (Figure 12.12).

It should be noted that unilateral micropthalmia is relatively common in C57Bl/6 stains, more frequently affecting females than males and the right eye more commonly than the left (Smith *et al.* 1994).

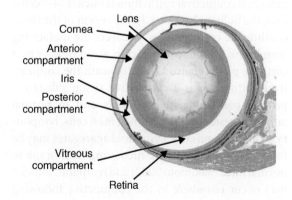

Figure 12.12   Overview of major components of eye structure.

12.4.1   *Eyelids*

The eyelids of mice are not routinely examined as part of most sampling protocols. It is, however, important to be aware of their morphology in case fragments of tissue appear on a slide or in the event of a phenotype affecting them. On the outer surface the eyelids are covered by haired skin comprising keratinized stratified squamous epithelium composed of 7–12 layers of keratinocytes and hair follicles and associated structures similar to those found elsewhere. Just before the mucocutaneous junction there is a row of cilia or eyelashes, which, although apparently longer, are in other respects morphologically similar to pelage hairs (Tauchi *et al.* 2010). The branching stratified epithelial lined ducts of the Meibomian glands open on to the eyelid surface at the mucocutaneous junction (Figure 12.11 and 12.13). The Meibomian glands consist of lobules of sebaceous epithelium similar to the sebaceous glands of hair follicles. The eyelid stroma consists of a connective tissue containing a dense collagenous plate and striated muscle.

At the mucocutaneous junction, the epithelium abruptly changes to a thinner stratified but nonkeratinized epithelium, which makes up the palbebral conjunctiva. Goblet cells are also found with varying densities in this epithelium and can be visualized with Alcian blue/PAS stains. Around most of the globe the conjunctival epithelium folds back at the fornix to cover the sclera (bulbar conjunctiva). The palbebral conjuctival epithelium is thicker (4-7 cells) than the bulbar (2-4 cells). In the region of the inner canthus the conjunctiva also covers the nictatating membrane, a tissue fold which is given rigidity by a cartilage plate (Figure 12.13). Beneath the conjunctival epithelium is a connective tissue substantia propria containing blood vessels and in normal animals a resident population of mast cells, lymphocytes and plasma cells. Lymphoid aggregates may be found associated with the nictitating membrane in normal mice (Sakimoto 2002) and lymphoid follicles may occur elsewhere in the conjunctiva following antigen stimulation.

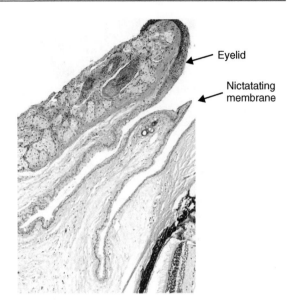

Figure 12.13   Eyelid from inner canthus with nictitating membrane.

12.4.2   *Sclera and cornea*

Although the sclera in mice is structurally tough, it is sufficiently thin that the pigmented choroid may be seen through it giving the globe a dark appearance in pigmented strains. Microscopically, the sclera is composed of three connective tissue layers. The outer layer or episclera is composed of fibrovascular tissue and forms points of attachment for the ocular muscles. The episclera is lined by a dense collagenous stroma, which merges with a thin inner layer of collagen (lamina fusca) that loosely attaches the stroma to the underlying choroid. The stroma contains haphazardly arranged collagen fibres and occasional fibroblasts. Emissarial canals traverse the sclera and contain nerves and blood vessels.

The sclera joins the cornea abruptly at the limbus. The surface of the cornea is lined by a nonkeratinizing stratified squamous epithelium, which usually appears to be 4–7 cells thick with the light microscope (although up to 13 layers may be seen using electron microscopy Henriksson *et al.* 2009) (Figure 12.14) and is thicker centrally than at the periphery. The basal cells are cuboidal and, in

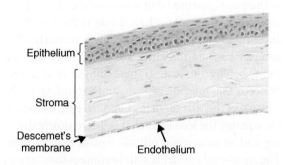

Figure 12.14   Layers of cornea.

normal animals, few, if any mitoses are present and these should be found in this layer only. Mitoses in higher layers suggest an increase in proliferation. The epithelium is attached to a basal lamina which can be stained with PAS (Figure 12.6). Subepithelial mineralization in the cornea is a frequent finding in ageing mice (Figure 12.15). Mineralization can be confirmed by using special stains such as Von Kossa's.

There is generally considered to be no mouse equivalent to Bowman's layer (anterior limiting lamina) in the primate eye although this is disputed (Henriksson *et al.* 2009) and a distinct layer can be demonstrated by electron microscopy. Beneath the epithelial surface the bulk of the cornea is made up of a connective tissue stroma composed of specialized fibroblasts (keratocytes) with elongated

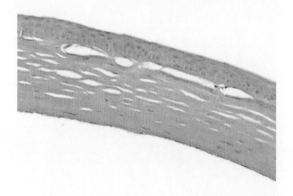

Figure 12.15   Mineralization of the subepithelial layer of the cornea is not uncommon in ageing mice.

oval nuclei scattered between a matrix of organized lamellae (or layers) of collagen fibres (principally types I and V collagen) and proteoglycans (Hassell and Birk 2010).

The inner surface of the cornea is lined by a single cuboidal endothelial (mesothelial) cell layer which secretes an acellular layer composed of collagen types IV, VIII, XVIII, proteoglycan and other matrix proteins (Descemet's membrane). Descemet's membrane has an anterior banded zone, which is present at birth and reaches maximum thickness by 6 weeks of age, and a posterior nonbanded zone, which is secreted by the endothelial cells and increases in thickness from 4 to 24 months, leading to a gradual overall increase in thickness of this layer with age (Jun *et al.* 2006).

12.4.3   *Uvea*

The uvea consists of the iris, ciliary body and choroid (Figure 12.16).

The iris extends from the junction of the ciliary body and choroid. The iris is covered on its anterior surface by the anterior border layer, which is actually a condensation of the underlying stroma and on its posterior surface by a double-layered, iris pigmented epithelium (IPE). The apices of the two layers of the IPE face each other and there are therefore two basal laminae facing the iris stroma and the lumen of the posterior chamber respectively. The IPE extends as the outer pigmented epithelial layer of the ciliary body. The core of the iris is made up of fibrovascular connective tissue with melanocytes,

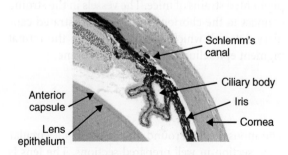

Figure 12.16   Iris, ciliary body, canal of Schlemm.

which contain pigmented melanosomes in pigmented mouse strains that can obscure any other cellular features. In nonpigmented mice, the melanasomes are still present, but lack pigment. Macrophages containing ingested melanin may also be present. The iris also contains fine collagen fibres, circular and radial smooth muscle fibres (which contain the visual pigment melanopsin, Xue *et al.* 2011) but these can be hard to visualize in standard histological preparations in the mouse.

The ciliary body is formed by several finger-like projections, consisting of a collagneous stroma, which contains smooth muscle fibres covered by a double layer of epithelial cells. The epithelial layer closest to the stroma is continuous with the retinal pigment epithelium and the IPE and the overlying nonpigmented epithelial layer (closest to the vitreous) is continuous with the retina (Smith *et al.* 2002c). As in the iris, the apices of the two epithelial layers face each other. The nonpigmented layer may be the major source of fibrillin making up the zonules.

The canal of Schlemm is an endothelial-lined vessel that drains aqueous humour into the venous circulation. It is found at the junction between the cornea and iris and is often indistinct in standard sections of the mouse eye. The entrance to the canal is bounded by a meshwork of endothelial-lined collagen trabeculae (Smith *et al.* 2002c).

The choroid is a relatively thin layer that provides the blood supply to the retina and is located between the lamina fusca of the sclera and the retinal pigment epithelium. Beneath the lamina fusca is a highly vascular stroma that contains melanocytes in pigmented strains of mice. The vessels in the stroma connect to the choriocapillaris – a fenestrated capillary network which is separated from the retinal pigment epithelium by Bruch's membrane.

12.4.4 Lens

The mouse lens is round and is seen as a circular cross section in well prepared sections. The lens is contained within a collagen and carbohydrate rich capsule that stains positively with PAS (Figure 12.6). The lens is held in place by zonules, which attach to the nonpigmented epithelium of the ciliary body and the lens capsule. The capsule is thicker at the anterior aspect of the lens than posteriorly. The anterior capsule thickness increases with age and varies between mouse strains (Danysh *et al.* 2008). Beneath the anterior capsule and extending to the equator of the lens there is a cuboidal epithelium. At the equator the lens epithelial cells are slowly dividing and so occasional mitotic figures may be seen. The equatorial epithelial cells become columnar and start to elongate to form the lens fibres that make up the outer cortex of the lens. Nuclei of the elongating fibres can be seen as the nuclear bow (Figure 12.17). Disruption to the ordered appearance of nuclei in this region can indicate abnormalities of lens function although it is important to be ensure that the lens has been correctly orientated so that artefactual distortions are not present. As the lens fibres age and migrate towards the medulla of the lens they degenerate and organelles and nuclei are lost leading to an 'organelle free zone' in the inner medulla (Bassnett 2009). With age, more nuclei and nuclear remnants may be retained and may be seen scattered throughout the lens which may contribute to increased opacity (Pendergrass *et al.* 2011).

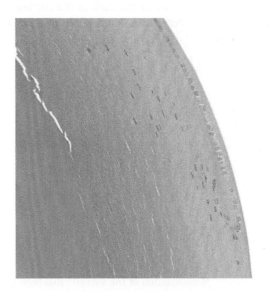

Figure 12.17 Lens fibre nuclei in regular arrangement at nuclear bow. Artefactual cracking of the lens can be seen and is hard to avoid in routine sections.

Age-related cataracts are a common spontaneous finding in old mice (Wolf *et al.* 2000) (Figure 12.9).

12.4.5 *Vitreous*

The vitreous lies between the posterior lens and the retina. In routine haematoxylin and eosin stained sections the vitreous may not be visible as it is composed of a highly hydrated, extracellular matrix (Bishop *et al.* 2002), which tends to be removed during processing. Remnants of hyaloid blood vessels may occasionally be observed in mice as in other species especially in young animals (Smith *et al.* 2002d) (Figure 12.1) and macrophage-like hyalocytes may also occasionally be observed. On electron microscopic examination, fine collagenous attachments to the inner limiting membrane of the retina may be noted (Smith *et al.* 2002d).

12.4.6 *Retina*

The retina is the light-sensitive organ of the eye and is separated from the choroid by Bruch's membrane, which consists of a layer of collagen and elastic fibres sandwiched between the basement membranes of the RPE and the choroid capillaries (Figure 12.18). The RPE consists of a single layer of flattened epithelial cells containing melanosomes, which contain pigment in pigmented strains of mice but are unpigmented in albino strains. Tight (occludens) and adherens junctions between both the RPE cells and the endothelial cells of the choroidal blood vessels form the basis of the blood-retinal barrier (Runkle and Antonetti 2011).

The light sensitive retina is composed of nine highly ordered layers (Figure 12.18). The internal surface of the retina is lined by a PAS positive basement membrane (inner limiting membrane) made up of the end processes of a type of glial cell, the Muller cells, whose cell bodies are found in the inner nuclear cell layer and also have processes that project to the outer limiting membrane where they form tight junctions with the photoreceptor cells.

Thinning, loss or disorganization of the cellular layers of the retina suggests atrophy and it is important to be aware that certain strains of mice,

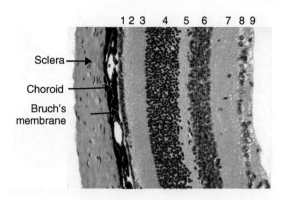

Figure 12.18 High-power photomicrograph showing normal layers of the retina and choroid. Retinal layers 1 = retinal pigment epithelium, 2= outer segment of photoreceptors, 3= inner segment of photoreceptors, 4= outer nuclear layer, 5 = outer plexiform layer, 6 = inner nuclear layer, 7 = inner plexiform layer, 8 = ganglion cell layer, 9 = nerve fibres.

particularly those based on C3H and FVB stock, have a genetic predisposition to retinal degeneration due to carrying the *Pde6b*[rd1] gene. Light-induced retinal atrophy is common particularly in albino strains (Figure 12.19). Retinal atrophy is characterized by variable loss of the photoreceptor layers, outer nuclear and outer plexiform layers (Taylor 2011). Occasional nuclei or nuclear debris may be seen in the photoreceptor layer as a background observation but increased numbers of nuclei particularly in association with vacuolation of the photoreceptor cells or changes in the RPE will increase the suspicion of a pathological change.

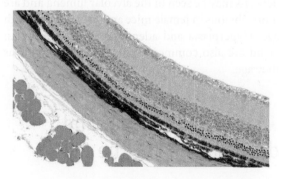

Figure 12.19 Example of retinal atrophy with partial loss of outer nuclear layer.

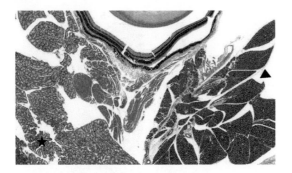

Figure 12.20 Harderian (star) and intraorbital lacrimal gland (arrow head).

12.4.7 *Intraorbital glands*

Behind the eye in the orbit there are two glandular structures in the mouse, the Harderian gland and the intraorbital lacrimal gland (Figure 12.20). The mouse also has an extraorbital lacrimal gland, which is located at the base of the ear adjacent to the parotid salivary gland. The Harderian gland is a tubular-alveolar gland that empties via a single excretory duct onto the nictitating membrane (Cohn 1955). The ductal system is lined by a low stratified epithelium, but may be hard to appreciate in routine sections. The epithelial cells that make up the alveoli are columnar with basally located nuclei and vacuolated cytoplasm. Although there are two types of epithelial cell described, these cannot be separated easily with the light microscope. Type A cells contain lipid vacuoles and porphyrin (which will autoflouresce) and type B cells contain smaller lipid vacuoles. Alveoli are surrounded by myoepithial cells and a sparse stroma. Porphyrin pigment deposits may be seen in the alveolar lumena and are more obvious in female mice and may increase with age. Hyperplasia and adenomas of the Harderian gland are also common in many mouse strains as they age.

12.5 Background and development of the ear

The main features of the mouse ear are similar to most nonprimate mammalian species in which the middle and inner ear are contained within a large auditory bulla rather than the temporal bone. The ear is divided into three major anatomical compartments: the external, middle and inner ear (Figure 12.21).

Embryologically, the inner ear develops from a thickening of the surface ectoderm of the hind brain – the otic placode. The otic placode starts to invaginate forming first the otic pit and then the otic vesicle. The otic vesicle elongates and differentiates to form the cochlear duct, semicircular canals, uticle, saccule and endolympatic duct. Mesenchymal cells around the otic vesicle condense to form the otic capsule. The middle ear develops from the first pharyngeal pouch and the ear bones from the first and second branchial arches (Mallo 1998). The external acoustic meatus of the outer ear is

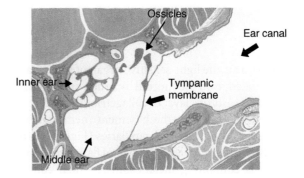

Figure 12.21 Illustration of the major compartments of the ear.

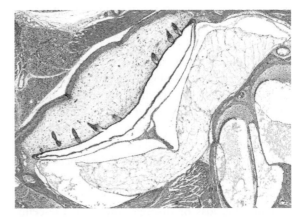

Figure 12.22 Middle ear of mouse is still developing at postnatal day 12.

closed by an epithelial plug at birth opening, 3 to 5 days postnatally. The tympanic membrane and pinna also continue to develop after birth reaching their adult size at approximately 18 days and 4 weeks after birth respectively (Kobayashi *et al.* 2011) (Fig. 12.22).

12.6   Sampling technique for the ear and associated structures

The ear pinna can be examined from routinely fixed sections and is usually sectioned along its length.

The middle and inner ears, which are enclosed in the auditory bulla in mice, are generally not examined routinely and require specialist histological preparations to examine in detail. These specialist techniques are beyond the scope of this book and are therefore referenced briefly below for guidance. Increasing 3D reconstructive imaging techniques, including CT and MRI, are being used for analysis of the inner ears of mice (Rau *et al.* 2012).

Standard immersion fixation in formalin followed by decalcification for 1–2 days in formic acid can be sufficient to aid with the diagnostic investigation of relatively common clinical presentations such as head tilt. Head tilt may be due to CNS lesions but is commonly related to middle-ear pathology in mice including otitis media (Figure 12.23) or polyarteritis secondary to glomerulonephropathy (Glaister 1986). To obtain diagnostic sections of Zymbals gland and the middle ear, after fixation and decalcification, the skin and lower jaw of the head should be removed and a block of tissue sectioned from either side of the external ear canal opening. The resultant block of tissue should be embedded cranial side down and step sections made through the tissue until the relevant tissues can be identified in the section.

The sensory and balance organs of the inner ear are susceptible to drug- and chemical-induced pathology and may be altered or lost in specific strains of genetically modified mice. For investigation of the inner ear, immersion fixation alone

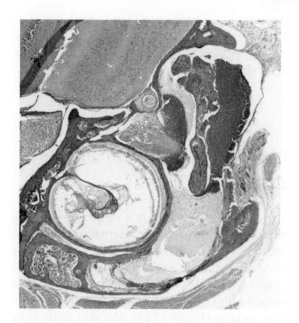

Figure 12.23   Otitis media.

is generally not considered adequate (Forge at al 2011). Methods that involve flushing (direct perfusion) the auditory bulla and cochlea with fixative or whole body perfusion (Bohne *et al.* 2001) are required to obtain optimal fixation. Formalin or paraformaldehye are adequate for most light microscopy techniques whereas glutaraldehyde should be used for electron microscopy. Direct infusion of fixative is considered to give the best morphology and can be achieved by dissecting the auditory bulla from the skull, opening the cochlea at both ends by making a small opening at the apex and opening the base of by removing the round window and removal of the small stapes bone from the oval window and then carefully injecting fixative. Once fixed, the cochlea can then be examined using whole mount protocols or by routine histology (Forge *et al.* 2009). Whole-mount methods may be most appropriate for detailed examination of the neuroepithelium of the inner ear but are beyond the scope of this text (see Forge *et al.* 2009 for an overview). Sections of the cochlea can be made following processing and embedding of the dissected inner ear in plastic resins (Bohne *et al.* 2001) or following decalcification of the preparation for 5 days in EDTA and embedding in paraffin wax (van Spaendonck *et al.* 2000). Tissue can be embedded to make longitudinal sections parallel to the central core (modiolus) of the cochlea.

12.7 Anatomy and histology of the ear and associated glands

The external ear consists of a simple, upright flap or pinna that surrounds the external auditory meatus or ear canal (which extends to the tympanic membrane) and the associated Zymbal's gland. The pinna is stiffened by a central cartilage core covered distally with keratinized squamous, haired skin on both surfaces (Figure 12.24). The skin tends to be slightly thicker than that seen on the trunk of the mouse. Towards the base of the ear canal the internal surface epithelium is nonhaired keratinized and squamous. Sebaceous glands are relatively abundant and present in the dermis on the inner and outer surfaces. Laterally, at the base of the pinna, there is a complex sebaceous gland (Zymbal's gland). The gland is made up of typical sebaceous epithelial cells with abundant foamy eosinophilic cytoplasm arranged in acini which empty via keratinized squamous ducts onto the surface of the external auditory meatus (Figure 12.25).

The tympanic membrane is very thin in mice and consists of three layers (Figure 12.26). The external layer of keratinized squamous epithelium, which is continuous with the lining of the ear canal, covers a thin collagen core. The inner surface is covered by respiratory epithelium which lines the middle ear and auditory (Eustachian) tube, which links the ear and nasopharynx (Figure 12.27).

The middle ear consists of an air-filled cavity behind the tympanic membrane, which contains the ossicles and the entrance to the auditory tube. As in other species, the mouse has three ossicle bones. The malleus attaches to the tympanic membrane at one end and is joined to the incus via a cartilagenous synchondrosis. The incus and stapes are joined by a

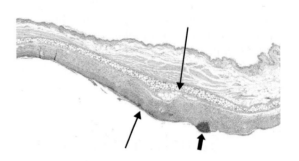

Figure 12.24 Pinna of mouse with hyperplasia (thin arrow) and focal accumulation of neutrophils (short arrow) in the epidermis secondary to self-trauma. Central cartilage indicated by long arrow.

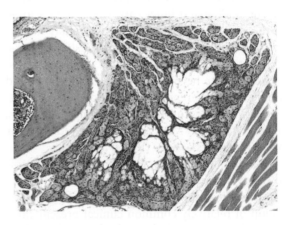

Figure 12.25 Zymbal's gland.

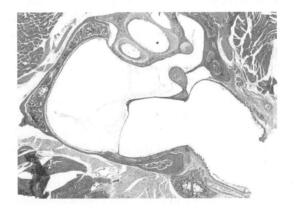

Figure 12.26 Middle ear.

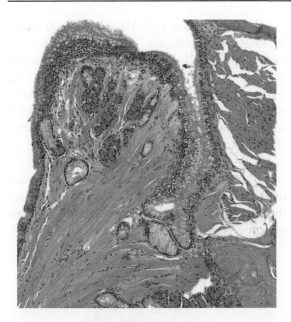

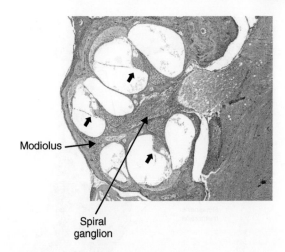

Modiolus

Spiral
ganglion

Figure 12.28   Longitudinal section of the cochlea with central modiolus and three organs of Corti (thick arrows).

Figure 12.27   Entrance to Eustachian tube from nasopharynx, lined by numerous goblet cells.

true joint and the stapes attaches to the oval window of the cochlea (Figure 12.26).

The inner ear consists of a series of bony canals that contain the cochlea (Figure 12.28), vestibule and semicircular canals of the vestibular apparatus. Within the canals are membrane-bound ducts, which are separated from the surrounding bone by extracellular fluid (perilymph). Perilymph is thus found in two compartments, the scala tympani, which ends at the round window and follows the outer curvature of the cochlea, and the scala vestibule, which ends at the oval window and follows the inner curvature of the cochlea.

For a detailed review of the cellular and molecular structure of the inner ear see Forge and Wright (2002). The major histological features are summarized below. Within the cochlea the membrane-bound compartment is triangular in cross-section, contains endolymph and is known as the cochlear duct or scala media. The three walls that make up the triangle have distinct morphological and functional features. The basal wall contains the organ of Corti (Figure 12.29), which is the main sensory organ of hearing. The organ of Corti is made up of inner and outer hair cells, which are distinguished by a prominent microvillus border on their apical surfaces, separated by supporting cells creating a mosaic of epithelial cells. There is a single row of inner hair cells which are closest to the modiolus and three or four rows of outer hair cells (Figure 12.29). Inner hair cells are rounded and flask shaped, being wider basally and outer hair cells are elongated and cylindrical. Outer hair cells increase in length from the base to the apex of the cochlea while the inner hair cells do not vary in length. Nerve cell endings synapse with the base of the hair cells, with cell bodies in the spiral ganglion. Supporting cells are columnar epithelial cells with basal nuclei and are anchored to the underlying basilar membrane, which consists predominantly of type IV collagen, laminin, fibronectin and an ear-specific protein, usherin. There are several types of supporting cells, inner and outer pillar cells separate the inner and outer hair cells, Hensen's cells, which contain lipid droplets, lie outside the outer hair cells, and Deiters' cells, which support the outer hair cells. Deiters' cells are attached to the basement membrane and support the basal pole of the outer hair cells. Cytoplasmic extensions from Deiters cells extend to the surface of the epithelium and join the cell membrane of the outer hair cells at the latero-apical membrane. The lateral membranes are not connected and the

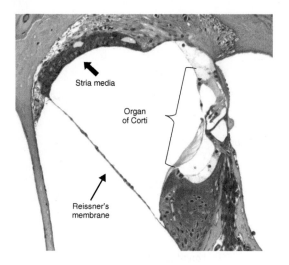

Figure 12.29   Resin embedded semithin sections can provide more detail of inner ear structures. Toluidine blue
stain.

The organ of Corti and stria vascularis are joined by Reissner's membrane, which forms the third edge of the triangle. Reissner's membrane consists of a single layer of flattened mesothelial cells lining the scala vestibule, separated by a basement membrane from a single layer of low columnar epithelial cells that line the scala media (Pace *et al.* 2001) (Figure 12.29).

In the vestibule and semicircular canals there are a further five foci of sensory cells, which are important in sensing balance. These structures are the maculae in the utricle and saccule of the vestibule and the cristae ampullaris in the semicircular canals

resultant space between supporting cells and outer hair cells is filled with extracellular fluid (space of Nuel). The inner and outer hair cells are separated by a fluid-filled space (tunnel of Corti). The apical surface of the sensory epithelium is covered by the acellular, gelatinous tectorial membrane, which consists of collagen types II, V and IX and a number of glycoproteins unique to the inner ear, including otoglein and tectorin.

Laterally, the scala media is composed of the stria vascularis, which is attached to the fibrocytic spiral ligament. The stria vascularis consists of a stratified epithelium that surrounds a dense capillary bed. The epithelium consists of three major cell types: marginal, intermediate and basal. Marginal cells form a single layer of cuboidal epithelium lining the lumenal surface and interdigitate basally with the intermediate cells. The intermediate cells produce melanin, and are sometimes referred to as melanocytes, in pigmented strains. Melanin granules can be exported from the intermediate cells and taken up by the marginal and basal cells. There are two to three layers of basal cells, which are flattened epithelial cells that line the spiral ligament (Raphael and Altschuler 2003).

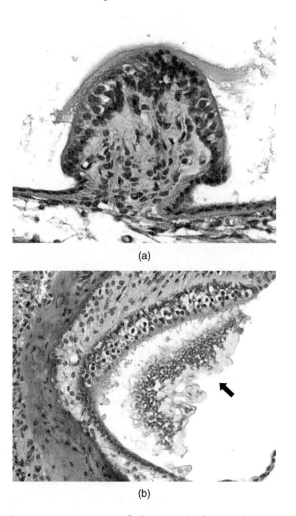

(a)

(b)

Figure 12.30   Organs of the semicircular canals – (a) ampulla and (b) macula (otoconia indicated by arrow).

(Figure 12.30). These sensory regions are similar to the organ of Corti being constructed of hair cells and supporting cells with an overlying acellular structure. There are two types of hair cells that can be distinguished ultrastructurally by differences in the length of cilia, but cannot be distinguished by standard light microscopy techniques. The cristae ampullaris are made up of a layer of sensory hair cells overlying a central projecting cone of connective tissue. An acellular layer known as the cupula covers the surface of the hair cells. The maculae, by contrast, are flat foci of sensory hair cells with basophilic granular concretions of calcium carbonate known as otoconia on their surface (Li *et al.* 2008).

Dark cell regions around the cristae and the macula of the utricle are the functional equivalent of the stria vascularis and consist of a luminal layer of dark cells (equivalent of the marginal cells) overlying a population of pigmented cells.

The remaining membranous structures of the vestibule and semicircular canals are lined by non-specific low cuboidal mesothelium.

References

Bassnett, S. (2009) On the mechanism of organelle degradation in the vertebrate lens. *Experimental Eye Research* **88**, 133–139.

Bishop, P.N., Takanosu, M., Le Goff, M. and Mayne, R. (2002) The role of the posterior ciliary body in the biosynthesis of vitreous humour. *Eye (London)* **16**, 454–460.

Bohne, B.A., Harding, G.W. and Ou, H.C. (2001) Preparation and evaluation of the mouse temporal bone, in *Handbook of Mouse Auditory Research: From Behavior to Molecular Biology* (ed. J.F. Willott), CRC Press, Boca Raton, FL.

Cohn, S.A. (1955) Histochemical observations on the Harderian gland of the Albino mouse. *Journal of Histochemistry and Cytochemistry* **3**, 342–353.

Danysh, B.P., Czymmek, K.J., Olurin, P.T. *et al.* (2008) Contributions of mouse genetic background and age on anterior lens capsule thickness. *Anatomical Record (Hoboken)* **291** (12), 1619–1627.

Dubielzig, R.R., Ketring, K., McLellan, G.J. *et al.* (2010) *Veterinary Ocular Pathology*, Saunders, Edinburgh.

Escher, P. and Schorderet, D.F. (2011) Exploration of the visual system: part 1: dissection of the mouse eye for RNA, protein and histological analysis. *Current Protocols in Mouse Biology* **1**, 445–462.

Forge, A., Taylor, R. and Harpur, E.S. (2009) Ototoxicity, in General and Applied Toxicology (eds B. Ballantye, T. Marrs and T. Syversen), Wiley-Blackwell, Hoboken, NJ.

Forge, A., Taylor, R. and Bolon, B. (2011) Toxicological neuropathology of the ear, in Fundamental Neuropathology for Pathologists and Toxicologists: Principles and Techniques (eds B. Bolon and M.T. Butt), John Wiley & Sons, Hoboken, NJ.

Forge, A. and Wright, T. (2002) The molecular architecture of the inner ear. *British Medical Bulletin* **63**, 5–24.

Glaister, J. (1986) *Principles of Toxicological Pathology*, Taylor & Francis, London.

Hassell, J.R. and Birk, D.E. (2010) The molecular basis of corneal transparency. *Experimental Eye Research* **91**, 326–335.

Heimann, M., Kasermann, H.P., Pfister, R. *et al.* (2009) Blood collection from the sublingual vein in mice and hamsters: a suitable alternative to retrobulbar technique that provides large volumes and minimizes tissue damage. *Laboratory Animals* **43**, 255–260.

Henriksson, J.T., McDermott, A.M. and Bergmanson, J.P.G. (2009) Dimensions and morphology of the cornea in three strains of mice. *Investigative Ophthalmology and Visual Science* **50**, 3648–3654.

Jun, J.A., Chakravarti, S., Edelhauser, H.F. and Kimos, M. (2006) Aging changes of mouse corneal endothelium and Descemet's membrane. *Experimental Eye Research* **83**, 890–896.

Kaufmann, M.H. and Bard, B.L. (1999) *The Anatomical Basis of Mouse Development*, Academic Press, San Diego, CA.

Kobayashi, S., Takebe, T., Zheng, Y.-W. *et al.* (2011) Presence of cartilage stem / progenitor cells in adult mice auricular perichondrium. *PLoS ONE* **6** (10), e26393. doi:10.1371/journal.pone.0026393.

Latendresse, J.R., Warbrittion, A.R., Jonassen, H. and Creasy, D.M. (2002) Fixation of testes and eyes using a modified Davidson's fluid: comparison with Bouin's fluid and conventional Davidson's fluid. *Toxicologic Pathology* **30**, 524–533.

Li, A., Xue, J. and Peterson, E.H. (2008) Architecture of the mouse utricle: macular organisation and hair bundle heights. *Journal of Neuro Physiology* **99**, 718–733.

Mallo, M. (1998) Embryological and genetic aspects of middle ear development. *International Journal of Developmental Biology* **42**, 11–22.

Morawietz, G., Ruel-Fehlert, C., Kittel, B. *et al.* (2004) Revised guides for organ sampling and trimming in rats and mice – part 3. *Experimental and Toxicologic Pathology* **55**, 433–449.

Pace, A.J., Madden, V.J., Henson, O.W. Jr *et al.* (2001) Ultrastructure of the inner ear of NKCC1-deficient mice. *Hearing Research* **156**, 17–30.

Pendergrass, W., Zitnik, G., Urfer, S.R. and Wolf, N. (2011) Age-related retention of fiber cell nuclei and nuclear fragments in the lens cortices of multiple species. *Molecular Vision* **17**, 2672–2684.

Ramos, M., Reilly, C.M. and Bolon, B. (2011) Toxicologic pathology of the retina and optic nerve, in *Fundamental Neuropathology for Pathologists and Toxicologists: Principles and Techniques* (ed. B. Bolon and M.T. Butt), John Wiley & Sons, Hoboken, NJ.

Raphael, Y. and Altschuler, R.A. (2003) Structure and innervation of the cochlea. *Brain Research Bulletin* **60**, 397–422.

Rau, C., Hwang, M., Lee, W.-K. and Richter C.-P. (2012) Quantitative X-ray tomography of the mouse cochlea. *PLoS ONE* **7** (4), e33568. doi:10.1371/journal.pone.0033568.

Runkle, E.A. and Antonetti, D.A. (2011) The blood-retinal barrier: structure and functional significance. *Methods in Molecular Biology* **686**, 133–148.

Sakimoto, T., Shoji, J., Inada, N. *et al.* (2002) Histological study of conjunctiva-associated lymphoid tissue in mouse. *Japanese Journal of Ophthalmology* **46**, 364–369.

Smith, R.S., Roderick, T.H. and Sundberg, J.P. (1994) Microphthalmia and associated abnormalities in inbred black mice. *Laboratory Animal Science* **44**, 551–560.

Smith, R.S., Hawes, N.L., Miller, J. *et al.* (2002a) Photography and necropsy, in *Systematic Evaluation of the Mouse Eye* (ed. R.S. Smith), CRC Press, Boca Raton, pp. 251–264.

Smith, R.S., John, S.W.M. and Nishina, P.M. (2002b) Posterior segment and orbit, in *Systematic Evaluation of the Mouse Eye* (ed. R.S. Smith), CRC Press, Boca Raton, FL, pp. 25–44.

Smith, R.S., Kao, W. W.-Y. and John, S.W.M. (2002c) Ocular development, in *Systematic Evaluation of the Mouse Eye* (ed. R.S. Smith), CRC Press, Boca Raton, FL, pp. 45–63.

Smith, R.S., Sundberg, J.P. and John, S.W.M. (2002d) The anterior segment and ocular adnexae, in *Systematic Evaluation of the Mouse Eye* (ed. R.S. Smith), CRC Press, Boca Raton, pp. 3–21.

Szel, A., Rohlich, P., Mieziewska, K. *et al.* (1992) Spatial and temporal differences between the expression of short- and middle-wave sensitive cone pigments in the mouse retina: a developmental study. *Journal of Comparative Neurology* **331**, 564–577.

Tauchi, M., Fuchs, T.A., Kellenberger, A.J. *et al.* (2010) Characterization of an *in vivo* model for the study of eyelash biology and trichomegaly: mouse eyelash morphology, development, growth cycle, and anagen prolongation by bimatoprost. *British Journal of Dermatology* **162**, 1186–1197.

Taylor, I. (2011) Mouse, in *Background Lesions in Laboratory Animals; A Colour Atlas* (ed. E.F. McInnes), Saunders, Edinburgh, p. 65.

Thompson, H. and Tucker, S. (2013) Dual origin of the epithelium of the mammalian middle ear. *Science* **339**, 1453–1456.

Van Herck, H., Baumans, V., Van der Craats, N.R. *et al.* (1992) Histological changes in the orbital region of rats after orbital bleeding. *Laboratory Animal* **26**, 53–58.

Van Spaendonck, M.P., Cryns, K., Van de Heyning, P.H. *et al.* (2002) High resolution imaging of the mouse inner ear by microtomography: a new tool in inner ear research. *The Anatomical Record* **259**, 229–236.

Williams, D. (2002) Ocular disease in rats: a review. *Veterinary Opthalmology* **3**, 183–191.

Wolf, N.S., Li, Y., Pendergrass, W. *et al.* (2000) Normal mouse and rat strains as models for age-related cataract and the effect of caloric restriction on its development. *Experimental Eye Research* **70**, 683–692.

Xue, T., Do , M.T.H., Riccio, A. *et al.* (2011) Melanopsin signalling in mammalian iris and retina. *Nature* **479**, 67–73.

Chapter 13
Musculoskeletal system

Cheryl L. Scudamore
Mary Lyon Centre, MRC Harwell, UK

13.1   Background and development

The mouse musculoskeletal system is fundamentally similar to that of other mammals with variations of the skeleton to accommodate rodent-specific features such as the head adaptations to the continuously growing incisors and the large olfactory system. Musculoskeletal tissues arise from mesoderm, with the bone and tendon arising from the lateral plate mesoderm and the muscles from the somite mesoderm. Mineralization of bones occurs by ossification of a cartilage template in limb bones (endochondral) or direct ossification of a fibrous stroma (intramembranous) in the flat bones of the skull. Ossification starts around embryonic day 14 in the forelimb long bones and continues into postnatal life (Patton and Kaufman 1995) with most bones having ossification centres by postnatal day 7, with some exceptions such as the patella.

13.2   Sampling techniques

Routine screening of mice should include examination of the musculoskeletal system. A range of nonhistological techniques including micro- CT, X-ray analysis, DEXA analysis of bone mineral density, biochemical analysis for mineral metabolism, breaking strength testing and muscle biomarkers can be used to screen for, or investigate, suspected musculoskeletal phenotypes (Bassett *et al.* 2012). Differential staining of whole foetuses with Alizarin red for bone and Alcian blue for cartilage can be used to map alterations in early bone development (McLeod 1980). Behavioural observations such as gait analysis and grip strength may also be made, particularly when muscle phenotypes are suspected.

Where no musculoskeletal phenotype is expected, standardized sections of a selected long bone including a joint and an example of skeletal muscle may be sufficient for a phenotypic histopathological screen. The femur and stifle joint along with hind

A Practical Guide to the Histology of the Mouse, First Edition. Cheryl L. Scudamore.
© 2014 John Wiley & Sons, Ltd. Illustrations © Veterinary Path Illustrations, unless stated otherwise.
Published 2014 by John Wiley & Sons, Ltd. Companion Website: www.wiley.com/go/scudamore/mousehistology

limb muscles, for example the quadriceps, biceps femoris or gastrocnemius are commonly sampled in routine protocols.

Bone and skeletal muscle may also be evaluated incidentally in vertebral sections if the spinal cord is processed within the spinal column and in sections of sternum, which may be taken primarily for examination of bone marrow.

Formalin or paraformaldehyde are adequate fixatives for routine examination of bone and muscle. For investigations of muscle fibre typing, using histochemical analysis or muscle diameter morphometry, frozen tissue is usually required (Wang and Kernell 2001; Briguet *et al.* 2004). Samples containing bone generally require decalcification prior to processing and embedding in paraffin wax. Decalcification can be performed in mice using EDTA or formic acid for 14 or 2–3 days respectively. In some circumstances, for example to evaluate mineral density, it may be helpful to examine nondecalcified bone. Nondecalcified bone can be examined using frozen sections or sections made from tissue embedded in resin. Both approaches may be technically demanding and require nonroutine histological equipment (Fuchs *et al.* 2006).

A longitudinal section through the femur, stifle joint and proximal tibia is commonly used for evaluation of bone, growth plate, articular cartilage and bone marrow (Morawietz *et al.* 2004). Coronal sections through the stifle joint where the patella is embedded face down may be useful for a more detailed assessment and scoring of damage to cartilage in the stifle joint (van Valburg *et al.* 1996). Scoring systems for damage to the articular surface have been developed for mice (Glasson *et al.* 2010; McNulty *et al.* 2011) adapted from the Mankin system widely used to assess human joint tissue (Mankin *et al.* 1971).

For routine evaluation of skeletal muscle, standardized sections of the biceps femoris muscle may be trimmed in longitudinal and or transverse orientation. Longitudinal sections are useful to assess striation but transverse sections are generally more useful for observation of fibre diameter, variation in staining, vacuolation and location of nuclei (Morawietz *et al.* 2004). Other muscles may be examined

for specific purposes, for example, diaphragm and intercostal muscles are in constant motion and may be particularly sensitive to myotoxic agents. Other muscles consist of predominantly fast twitch fibres e.g. tibialis anterior or slow twitch fibre e.g. soleus and so may be examined if specific effects are expected in different fibre types (Augusto *et al.* 2004).

13.3 Anatomy and histology

13.3.1 *Bone*

The histology of the standard section of femur and stifle joint and skeletal muscle will be described with specific features relevant to standard sections of sternum and vertebrae. Descriptions of variations for individual bones or muscles are beyond the scope of this text. Long bones in mice, as in other species, consist of the epiphysis at the ends separated by the metaphysis from the shaft or diaphysis.

Examination of the bones involves an awareness of the normal features of the articular cartilage, cortical and medullary bone and growth plates of mice. The cortical bone in mature animals consists of compact bone, which is composed of mature bone cells (osteocytes) embedded in a mineralized, collagen matrix (Figure 13.1). The collagen fibres

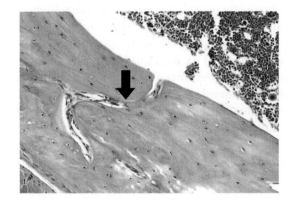

Figure 13.1 Mature compact bone from the femur showing osteocytes embedded in collagen matrix. Also shown is a nutrient artery (arrow) and bone marrow (top right).

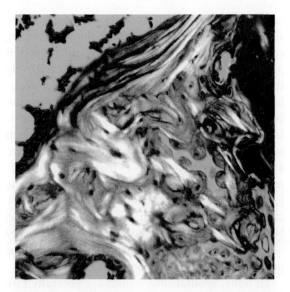

Figure 13.2   Collagen bundles demonstrated using polarized light.

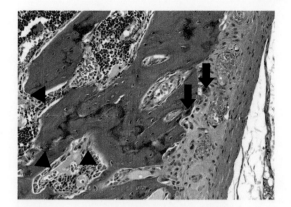

Figure 13.3   Bone showing osteoblasts (arrowheads) and multinucleated osteoclasts (arrows).

in mature compact (lamellar) bone are arranged in organized parallel pattern whereas in developing or repairing bone (woven) the collagen fibres are arranged haphazardly, which can be seen using polarizing light illumination (Figure 13.2). Osteoblasts, which are involved in the formation of bone, are plump cells with basophilic cytoplasm, which can be seen lining the medullary surface of the bone (Figure 13.3) and large, multinucleated osteoclasts, which are responsible for remodelling and removing bone, can be found lining trabeculae and are particularly numerous where the primary spongiosa is being remodelled (Figure 13.3).

The thickness of the articular cartilage in mice varies depending on the joint, but is generally 10 to 15 cells thick. The cartilage consists of four distinct layers of chondrocytes in lacunae embedded in an homogenous, pale eosinophilic collagen and proteoglycan matrix (Hughes *et al.* 2005). The surface or articular layer is two to three cells thick and the nuclei are flattened and parallel with the articular surface. In the intermediate or transitional layer the cells are rounder, often in pairs, and align at an angle to the articular surface (Figure 13.4). The deep or radial layer is the thickest layer of the mouse articular cartilage, the cells are larger and may be present in

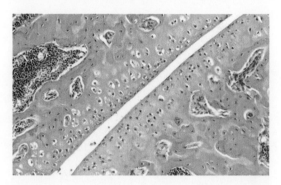

Figure 13.4   Articular surfaces of stifle joint - patella right and femur left.

small clumps aligned perpendicular to the surface. In the deepest layer adjacent to the subchondral bone, the matrix becomes calcified. Haematoxylin and eosin is an adequate stain for routine assessment and will allow identification of most changes e.g. cartilage necrosis and fibrillation. Cationic stains such as Toluidine Blue and Safranine-O that bind to proteoglycans can be useful in assessing more subtle damage in cartilage. Safranin-O, usually used with a Fast Green counterstain, stains glycosaminoglycans red and Toluidine Blue stains them dark blue or purple (Figure 13.5). Loss of this staining reaction indicates a decrease in glycosaminoglycans.

Arthropathy, or noninflammatory degenerative joint lesion, is common in the femoro-tibial joint of some mouse strains. The joint is usually enlarged and firm and microscopically there may be erosion

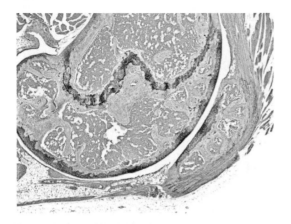

Figure 13.5   Toluidine blue can be used to stain cartilage in articular surfaces and joints.

Figure 13.6   Loss of femoral articular cartilage in case of arthropathy.

and loss of articular cartilage and proliferation of synovium and bone around the joint capsule (Figure 13.6).

The growth plates of mice of most strains do not usually completely close during the animal's life span but do reduce in size with age, although the long bones do not generally increase in length after about three months. Otherwise the main features of the growth plates are similar to other species. There are four zones of hyaline cartilage, which make up the growth plate. Hyaline cartilage generally consists of chondrocytes embedded in an extracellular matrix, which has a homogenous pale basophilic staining. From the epiphysis towards the medullary cavity these are the reserve zone, proliferating zone, hypertrophic zone and zone of calcification. The reserve or resting zone consists of chondrocytes, singly or in small groups, while in the proliferating zone the chondrocytes are flattened and aligned in parallel stacks. The zone of hypertrophy is relatively thin in mice and the cells are enlarged with a vacuolated appearance leading to an increased separation between nuclei. In the zone of calcification the cartilage becomes calcified and eventually separates into ossified trabeculae (primary spongiosa), which have a cartilage core (Figure 13.7) that is gradually remodelled by the action of osteoclasts to form the fully ossified secondary spongiosa of the medullary cavity (Figure 13.7). Haematopoietic tissue is found between the trabeculae of bone in the marrow cavity of most mice and persists through life although the proportion of adipose tissue gradually increases.

In bone, spontaneous lesions are generally rare but the tissue can be difficult to cut leading to common artefacts such as folding of tissue, 'chattering' and tearing (Figure 13.8 and Figure 13.12). Pyogranulomas may affect bones of the jaw and face and osteomyelitis may affect the feet, legs and tail in mice that have been group housed and where fighting has occurred. The lesions consist of aggregates of granulomas, which surround bacterial colonies,

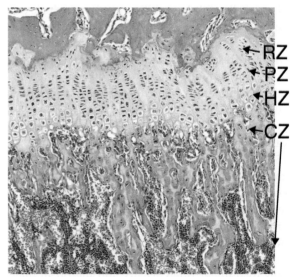

Figure 13.7   Growth plate of femur in 12-week-old mouse (RZ – reserve zone, PZ – proliferative zone, HZ – hypertrophic zone, CZ – calcification zone).

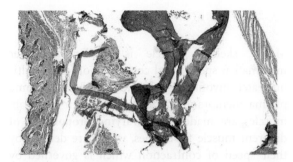

Figure 13.8   Bone can be hard to section and artefacts such as folding are common.

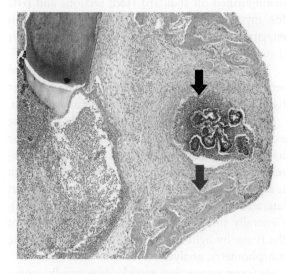

Figure 13.9   Pyogranuloma (black arrow) in jaw leading to disruption of bone (red arrows) and fibrosis.

Figure 13.10   Recent fracture with sharp edges of displaced bone, haemorrhage and acute inflammation.

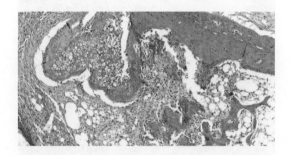

Figure 13.11   Chronic osteomyelitis in tail vertebrae secondary to traumatic damage.

often surrounded by fibrous tissue. There may be local destruction of bone (Figure 13.9). Accidental fractures of limbs and tail are also occasionally seen with acute lesions showing cleanly broken bone with haemorrhage and acute inflammatory infiltrates (Figure 13.10). Osteomyelitis can be seen as an extension of traumatic lesions. Acute lesions are characterized by a neutrophilic infiltrate and necrosis (Figure 13.11) with fibrosis and mononuclear cells predominating in older lesions.

The sternum of the mouse is routinely examined to evaluate the bone marrow. The sternum consists of the sternebrae bones (manubrium, sternebrae and xiphisternum) separated by fibrocartilagenous joints ending in the xiphoid cartilage (Figure 13.12).

The ribs and sternebrae join at the costochondral joints, which are also fibrocartilagenous. Degeneration (also known as chondropathy, aseptic necrosis or chondromucinous degeneration) of the intersternal or xyphoid cartilage can lead to deformity (commonly a prominent xiphoid) and can be seen in both young and older mice (Taylor 2012). The cause is not certain but the condition is degenerative with no associated inflammation. Histologically there is necrosis leading to areas where no chondrocytes are visible and the ground substance is fragmented or lost leaving a cystic space. Small clusters of chondrocytes may be present at the periphery.

Intervertebral disc structure is similar in mice to other species. The disc is attached to the vertebral bodies on either side by hyaline cartilage. The disc has two components an outer ring of dense

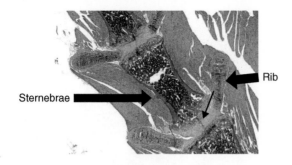

Figure 13.12 Sternum with ribs attached. Thin arrow shows artefactual folding in section.

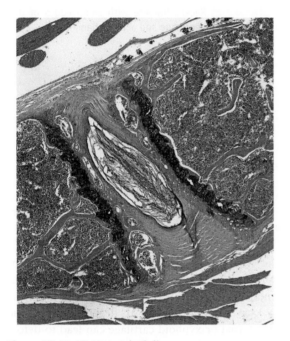

Figure 13.13 Intervertebral disc.

fibrocartilage the annulus fibrosus which consists of dense collagen fibres arranged in concentric lamellae and the core, the nucleus pulposus, which consists of a myxoid cartilaginous matrix (Figure 13.13). Both regions contain collagen and proteoglycans. The cell density in both the annulus and nucleus is very low with the chondrocytes in the annulus being thin and elongated and those in the nucleus being more rounded.

13.3.2 Muscle

Mouse skeletal muscle is similar to other mammalian species and composed of long, multi-nucleated myofibres surrounded by a sarcolemma plasma membrane. Depending on their function, muscles are made up of varying proportions of different muscle fibre types, which are defined by their speed of contraction which is governed by their ATPase activity and their expression of myosin heavy chain protein. These fibre types cannot be distinguished on standard H&E sections and IHC for myosin heavy isotypes or ATPase reaction enzyme histochemistry is needed to analyse their distribution (Kalmar *et al.* 2012).

Muscle can be evaluated in specifically selected sections but is also often found associated with other tissues, for example sections of femur and sternum.

Ideally, transverse sections of muscle should be examined but consistent transverse sections may be hard to obtain from complex muscles like gastrocnemius (which has two heads) and quadriceps, which is a complex of four muscles (rectus femoris, vastus lateralis, vastus intermedius and vastus medialis). Generally this does not complicate evaluation of the tissue by light microscopy but may complicate morphometric analysis.

In transverse section, muscle fibres usually appear polygonal with four to six small round nuclei beneath the sarcolemma for each fibre (Figure 13.14). In longitudinal section, the nuclei are elongated. On transverse section the cytoplasm of the myofibres is pale, eosinophilic and granular; in longitudinal sections regular striations are usually apparent that can be highlighted with special stains – for example PTAH (Figure 13.15). The cross-sectional area of each myofibre should be consistent within a muscle in normal animals; variations in size suggest an underlying pathology. Satellite cells are mononuclear cells lacking myofibrils, which are found around the basal membrane and act as stem cells for muscle regeneration. They may be difficult to distinguish from myofibre nuclei in routine sections.

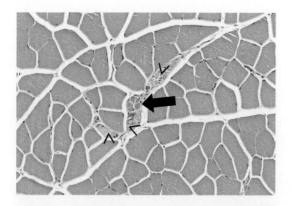

Figure 13.14   Transverse section of muscle showing muscle spindle in the centre (arrow). The small intrafusal fibres should not be confused with atrophic muscle fibres. Segments of nerve are also seen associated with the muscle spindle (arrow heads).

Myofibres are arranged in bundles separated by connective tissue perimysium (Figure 13.16, trichrome stain), which also contains blood vessels and nerve bundles.

Mechanoreceptor muscle spindles (Figure 13.14) may be seen in cross sections and the small, intrafusal fibres should not be confused with atrophic fibres. Muscle spindles are surrounded by a capsule of flattened modified fibroblasts (pavement cells). Within the capsule are two types of skeletal muscle intrafusal fibres – one to three large-diameter nuclear bag fibres and two to seven smaller nuclear chain fibres. The larger nuclear bag fibres have peripheral nuclei and the smaller nuclear chain fibres have central nuclei.

Central nuclei are uncommon in normal fibres apart from where the muscle fibres are close to a tendon insertion (Figure 13.17). Away from tendon insertions, centrally nucleated fibres may indicate muscle regeneration or other muscle pathology. Counting the number of central nuclei is sometimes used to assess muscle damage but can give false negative results when there is muscle fibre loss with replacement by fibrous tissue.

Vacuolation is also not seen in normal fibres but may be seen in certain pathologies or may be an

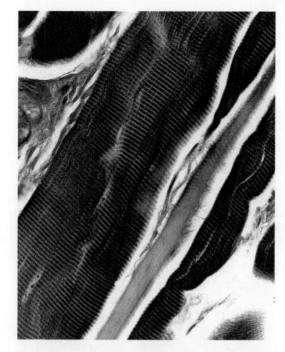

Figure 13.15   Muscle in longitudinal section stained with PTAH to demonstrate striations.

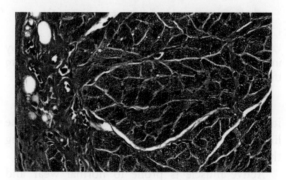

Figure 13.16   Massons trichrome stain demonstrates the connective tissue between individual muscle fibres, muscle bundles and around blood vessels in blue.

artefact particularly in frozen tissue. Contraction fibre artefact (Figure 13.18) is commonly seen in muscles of larger animals where the tissue has not been pinned out before fixation. This is less common in mouse tissue, even if the muscle is just immersed

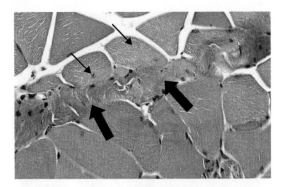

Figure 13.17 Central or internally placed nuclei (thin arrows) are uncommon in cross sections of normal muscle fibres except where the fibres are adjacent to connective tissue fascia (thick arrows) as seen in this image.

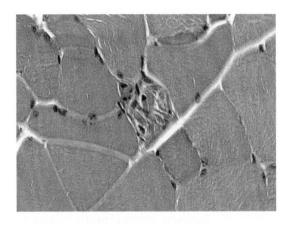

Figure 13.19 Single fibre necrosis start with fragmentation or vacuolation of the fibre, followed by infiltration with neutrophils and macrophages.

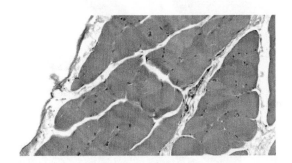

Figure 13.18 Contraction band artefact seen as wavy irregular more darkly eosinophilic bands perpendicular to the muscle fibre.

in fixative, but can be largely avoided by fixing the limbs whole after removal of the skin and dissecting the muscles for histology after fixation.

Active muscle is prone to normal wear and tear (including damage during handling of the animals) and therefore, in transverse sections, the appearance of single necrotic fibres is common. These fibres may be hypereosinophilic or paler than normal, fragmented and may have infiltrating macrophages, in transverse section striations are absent (Figure 13.19). Occasional smaller fibres with basophilic cytoplasm which represent regenerating fibres may also be seen. Increased numbers or groups of necrotic or basophilic fibres are suggestive of pathology.

References

Augusto, V., Padovani, C.R. and Campos, G.E.R. (2004) Skeletal muscle fiber types in C57B16J mice *Brazilian Journal of Morphological Sciences* **21**, 89–94.

Bassett, J.H.D., Gogakos, A., White, J.K. *et al.* (2012) Rapid-throughput skeletal phenotyping of 100 knockout mice identifies 9 new genes that determine bone strength. *PLoS Genetics* **8** (8), e1002858. doi:10.1371/journal.pgen.1002858.

Briguet, A. Courdier-Fruh, I., Foster, M. *et al.* (2004) Histological parameters for the quantitative assessment of muscular dystrophy in the mdf-mouse. *Neuromuscular Disorders* **14** (10), 675–682.

Fuchs, H., Lisse, T., Abe, K. and Hrabe de Angelis, M. (2006) Screening for bone and cartilage phenotypes in mice, in *Standards of Mouse Model Phenotyping* (ed. M. Hrabé de Angelis, P. Chambon and S. Brown), Wiley-VCH, Weinheim.

Glasson, S.S., Chambers, M.G., Van den Berg, W.B. and Little, C.B. (2010) The OARSI histopathology initiative – recommendations for histological assessments of osteoarthritis in the mouse. *Osteoarthritis Cartilage* **18** (suppl. 3), S17e23.

Hughes, L.C., Archer, C.W. and Gwynn, I. (2005). The ultrastructure of mouse articular cartilage: collagen orientation and implications for tissue functionality. A polarised light and electron microscope study and review. *European Cells and Materials* **9**, 68–84.

Kalmar, B., Blanco, G. and Greensmith, L. (2012) Determination of muscle fiber type in rodents. *Current Protocols in Mouse Biology* **2**, 231–243.

Mankin, H.J., Dorfman, H., Lippiello, L. and Zarins A. (1971) Biochemical and metabolic abnormalities in articular cartilage from osteo-arthritic human hips. II. Correlation of morphology with biochemical and metabolic data. *Journal of Bone and Joint Surgery American Volume* **53**, 523–537.

McLeod, M.J. (1980) Differential staining of cartilage and bone in whole mouse fetuses by alcian blue and alizarin red S. *Teratology* **22**, 299–301.

McNulty, M.A., Loeser, R.F., Davey, C. *et al.* (2011) A comprehensive histological assessment of osteoarthritis lesions in mice. *Cartilage* **2**, 354–363.

Morawietz, G., Ruehl-Fehlert, C., Kittel, B. *et al.* (2004) Revised guides for organ sampling and trimming in rats and mice – Part 3. A joint publication of the RITA and NACAD groups. *Experimental Toxicologic Pharmacology* **55**, 433–449.

Patton, J.T. and Kaufman, M.H. (1995) The timing of ossification of the limb bones, and growth rates of various long bones of the fore and hind limbs of the prenatal and early postnatal laboratory mouse. *Journal of Anatomy* **186**, 175–185.

Taylor, I. (2012) Mouse, in *Background Lesions in Laboratory Animals. A Color Atlas* (ed. E.F. McInnes), Saunders, London.

van Valburg A.A., van Osch, G.J., van der Kraan, P.M., van den Berg, W.B. (1996) Quantification of morphometric changes in murine experimental osteoarthritis using image analysis. *Rheumatology International* **15** (5), 181–187.

Wang, L.C. and Kernell, D. (2001) Fibre type regionalisation in lower hindlimb muscles of rabbit, rat and mouse, a comparative study. *Journal of Anatomy* **199**, 631–641.

Index

Indexer: Dr Laurence Errington

Illustrations are comprehensively referred to from the text. Therefore, significant items in illustrations (figures and tables) have only been given a page reference in the absence of their concomitant mention in the text referring to that illustration.

Printed and bound by CPI Group (UK) Ltd, Croydon, CR0 4YY

16/04/2025

14658459-0006